Deep Learning Applications

In Computer Vision, Signals and Networks

Deep Learning Applications

In Computer Vision, Signals and Networks

Edited by

Qi Xuan

Yun Xiang

Dongwei Xu

Zhejiang University of Technology, China

World Scientific

NEW JERSEY • LONDON • SINGAPORE • BEIJING • SHANGHAI • HONG KONG • TAIPEI • CHENNAI • TOKYO

Published by

World Scientific Publishing Co. Pte. Ltd.

5 Toh Tuck Link, Singapore 596224

USA office: 27 Warren Street, Suite 401-402, Hackensack, NJ 07601

UK office: 57 Shelton Street, Covent Garden, London WC2H 9HE

Library of Congress Cataloging-in-Publication Data

Names: Xuan, Qi, editor.

Title: Deep learning applications : in computer vision, signals and networks /
    edited by Qi Xuan, Yun Xiang, Dongwei Xu, Zhejiang University of Technology, China.

Description: New Jersey : World Scientific Publishing Co. Pte. Ltd., [2023] |
    Includes bibliographical references and index.

Identifiers: LCCN 2022040917 | ISBN 9789811266904 (hardcover) |
    ISBN 9789811266911 (ebook for institutions) | ISBN 9789811266928 (ebook for individuals)

Subjects: LCSH: Deep learning (Machine learning)

Classification: LCC Q325.73 .D45 2023 | DDC 006.3/1--dc23/eng20221209

LC record available at https://lccn.loc.gov/2022040917

British Library Cataloguing-in-Publication Data

A catalogue record for this book is available from the British Library.

For any available supplementary material, please visit
https://www.worldscientific.com/worldscibooks/10.1142/13158#t=suppl

Desk Editors: Jayanthi Muthuswamy/Amanda Yun

Typeset by Stallion Press
Email: enquiries@stallionpress.com

Preface

Deep learning, first introduced by Hinton *et al.* in 2006,[1] has brought great changes to the world. Modern deep learning technique relies upon deep neuron networks (DNN), which have a very long history and aim to mimic the functionality of human brains. After more than 15 years of development, it has become a great success. In many application areas, deep learning, for the first time in human history, has surpassed even human specialists. Naturally, people are eager to apply such a great and efficient tool to various application fields. The foremost application targets vision-based tasks. Historically, before the emergence of deep learning, no machine learning algorithm could match what the human eye and brain can do. Since deep learning has similar architecture as the human brain we expect it to have similar performance as humans. And it does! For the famous ImageNet contest, since the introduction of AlexNet,[2] deep learning algorithms have started to significantly outperform all the other machine learning algorithms in image classification and recognition. Therefore, they could have wide applications in real-world tasks. For example, it is difficult traditionally to recognize ship plates, identify the tiny deficiencies in industrious products, analyze the cop stress in agriculture, and estimate the environmental PM2.5 concentration using images. Unlike vision-based problems, signal processing usually deals with time-series data. The typical problems in signals include modulation, denoising, and classification. Moreover, besides vision and signal, existing techniques also face problems in the analysis of complex network structures. The internet connects most people in the world and forms a huge network. The behaviors and interactions of entities in a network at such a scale and complexity can hardly be efficiently and accurately analyzed. The emerging social networks are full of controversies. Etherum, a block-chain technology, is plagued by Ponzi schemes. In biology and chemistry, it is a very difficult task to predict the molecular biological activities. In general, it is important and urgent to develop corresponding DNN-based technologies for vision, signal, and network applications.

Researchers have devoted great enthusiasm and efforts to the development of DNNs. Currently, there is a surprisingly rich literature for your inspiration when searching for the solution of your problems. For computer vision tasks, it includes classification, object detection, and segmentation etc. He *et al.*[3] proposed ResNet, which enabled the training of very deep neural networks. Their model is widely used as the benchmark. Ren *et al.*[4] proposed faster R-CNN for object detection. Their model greatly improves the detection speed and shows great performance in the PASCAL visual object classes challenge. Chen *et al.*[5] proposed DeepLabV3+ for semantic segmentation. It employs the encoder-decoder structure and exploits multi-scale information using atrous convolution and spatial pyramid pooling. This technique can improve the accuracy of segmentation boundaries.

In signal processing, since signals are usually time series data, the deep learning method is mainly to deal with time series, such as pattern analysis, forecasting, segmentation, and classification, etc. Based on Long short-term memory (LSTM),[6] a series of deep learning models are developed for time series analysis and classification.[7] With the development of representation techniques, time series are also represented by other data forms. For example, Wang *et al.*[8] encoded time series as matrices with mathematical explanation. Lacasa *et al.*[9] established a graph-structured representation with geometric explanation. These techniques enable more diverse deep learning models, such as convolutional neural networks (CNN) and graph neural networks (CNN).

The rapid development of internet technology makes data acquisition easier. Many data platforms are gradually releasing large amounts for commercial development. These data have attracted many researchers to develop better algorithms. For example, Shang *et al.*[10] proposed an edge attention graph convolutional network (EAGCN) algorithm based on edge attention. It can evaluate the weight of each edge in the molecule via an edge attention layer for molecular networks. Hu *et al.*[11] presented a novel deep learning based scam detection framework using attention mechanism. For internet network data, TopoScope[12] used ensemble learning and Bayesian Network to reduce the observation bias, and reconstruct the internet topology by discovering hidden links. Existing techniques for conflict detection in social media mainly focus on macro-topic controversy detection.[13,14]

In this book, we mainly focus on the application of deep learning in vision, signal, and network. In the past few years, various deep learning models have been proposed to solve academic and industrial problems in the related areas. For real-world applications, the complexity of real-world scenarios and the restrictions on equipment cost limit the algorithm performance. Thus, this book will introduce the latest applications of deep learning. In particular, we focus on the following areas: computer vision, signal processing, and graph network.

Computer Vision aims to derive meaningful information from visual data, e.g., images and videos, to automate tasks or aid human decisions. For humans, vision accounts for the majority of our access to information and dominates our decisions. In some of the visual tasks, such as image classification, object detection, and behavior recognition, the neural network has outperformed humans.[15] Most computer vision algorithms are based on convolutional neural networks (CNNs). One of the most popular CNNs is ResNet,[3] which has solved the problem of training very deep neural networks and achieved state-of-the-art performance in the 2015 ImageNet Large Scale Visual Recognition Challenge.[16] Though many models are first proposed for image classification, they can also be used as powerful feature extractors. For example, in object detection, region-CNN[17] uses a CNN to extract features of latent regions where objects exist and then determines the category of the objects. Segmentation and classification are completely different tasks, but their models can share similar components.[18] In this book, these algorithms are evaluated on standard benchmark data.

In Chapter 1, Zhang, Chen, and Xiang introduce the method to estimate particle matter 2.5 (PM2.5). They deploy the sampling system and collect PM2.5 data and images for verification. The experimental results demonstrate the effectiveness and reliability of the method where feature maps and multi-view strategies are integrated.

In Chapter 2, Xu, Zhang, and Xiang introduce a two-stage end-to-end ship plate recognition technique based on R2CNN and ASTER. It is composed of text detection, rectification, and recognition techniques. They demonstrate the method in a real-world dataset collected from the port area. The method achieves high performance and real-time ship identification.

In Chapter 3, Shao, Chen, and Xuan introduce two bearing roller defect detection methods based on generative adversarial network (GAN) and object detection algorithms. The first method uses GAN to augment the dataset, followed by a ResNet classification network for defect detection. The second method uses the faster-RCNN model. If the two detection results are not matched, the DeepLabV3+ network is further used for small scratch detection. Experiments show that both methods can obtain excellent detection results.

In Chapter 4, Chen, Zhang, Xuan, and Xiang present the deep learning models on plant biotic and abiotic stresses. The application of deep learning on plant stress is not as extensive as in images. Plant stress-related researches have great potential.

Time series classification aims to classify data with time series properties. It is typically a pattern recognition task using supervised learning and is widely used in many areas such as radio communication, audio recognition, and medical data analysis, etc. For radio signal classification, the signals are identified for tasks such as modulation recognition,[19,20] radio frequency identification,[21] and channel estimation.[22] For audio recognition, there are many applications such as automatic speech recognition.[23] For medical data analysis, time series classification is typically used for ECG signal classification.[24] Deep learning-based time series classification can improve classification accuracy and the efficiency of model inference.

In Chapter 5, S. Gao, X. Gao, Zhou, Chen, Zheng, and Xuan introduce several commonly used pruning methods and further propose a mixed pruning technique for modulation recognition model. It can improve the compression rate of the model by combining filter-level and channel-level pruning methods. We also quantitatively evaluate the compressed models.

In Chapter 6, Feng, Chen, Zhou, Xu, and Xuan introduce the time series classification method based on broad learning system. With broad learning system deep learning models can be wider instead of deeper, to improve their efficiency while maintaining accuracy.

In Chapter 7, Yao, Zhou, Chen, Huang, Xu, and Xuan introduce a deep learning denoising method for radio modulation signals. It simplifies the signal denoising step and adaptively extracts high-level features. The authors combine GAN and signal properties to design an end-to-end model and further analyze the denoising performance based on constellation diagram, visualization, and time–frequency domain analysis.

In Chapter 8, Zhou, Qiu, Chen, Zheng, and Xuan expand the signal features in time domain and introduce an improved visibility graph (VG) for modulation recognition. The graph transforms the time series classification task into a graph classification one. The authors further design a multi-channel graph neural network framework that utilizes the expanded time-domain features to enhance the modulated signal recognition performance.

Many real-world systems can be naturally represented as **networks**, such as molecular network,[25] transaction networks,[26] internet networks,[27] and social networks.[28] Studying the structure of a network is an efficient way to understand and analyze the network. In molecular network, using network modeling methods to predict the biological activity of molecules is becoming popular, especially in areas such as drug development. For transaction networks, Ethereum has become one of the most popular block-chain platforms. The externally owned accounts (EOA), contract accounts (CA), the smart contract invocation, and assets transfer activities can be modeled as nodes and links of transaction networks. Fraud activities can be detected and identified through transaction pattern analysis.[29] The Internet is now composed of more than 70,000 autonomous systems (AS) and exhibits prominent complex structures.[30] Online social media have become important complex networks applications.[31] The online exchange of different opinions can lead to heated discussions and pollute the Internet environment. Social network analysis techniques based on graph mining can be applied for controversy detection[31] and maintain a more secure social ecosystem.

In Chapter 9, Peng, Zhang, Shu, Ruan, and Xuan introduce the AS classification method using graph convolutional networks (ASGNN) model. We demonstrate the effectiveness of the model for AS classification tasks and AS relationship features, enhancement.

In Chapter 10, Li and Zhang take both dynamic and static features for obtaining precise social network node embedding results. They develop a new feature calculation technique to integrate several node features, which significantly enhances the performance of controversy detection.

In Chapter 11, Jin, Zhou, Chen, Sheng, and Xuan construct a multidimensional graph neural network detection model for Ethereum Ponzi schemes detection. It employs the three main features, i.e., contract code, transactions, and network structure features.

In Chapter 12, Zhou, Tan, Zhang, and Zhao build a reliable prediction model for molecular biological activity prediction. It can be applied for molecular multi-feature fusion and adaptively fuses multiple features through the self-attention mechanism. The authors also use focal loss and gradient harmonizing mechanism (GHM) to analyze the imbalance between positive and negative samples in molecular biological activity data.

References

1. G. E. Hinton and R. R. Salakhutdinov, Reducing the dimensionality of data with neural networks, *Science*. **313**(5786), 504–507 (2006).
2. A. Krizhevsky, I. Sutskever, and G. E. Hinton, Imagenet classification with deep convolutional neural networks, *Communications of the ACM*. **60**(6), 84–90 (2017).
3. K. He, X. Zhang, S. Ren, and J. Sun, Deep residual learning for image recognition, in *2016 IEEE Conference on Computer Vision and Pattern Recognition (CVPR)*, 2016, pp. 770–778.
4. S. Ren, K. He, R. Girshick, and J. Sun, Faster R-CNN: Towards real-time object detection with region proposal networks, *Advances in Neural Information Processing Systems*. **28**, 1440–1448 (2015).
5. L.-C. Chen, Y. Zhu, G. Papandreou, F. Schroff, and H. Adam, Encoder-decoder with atrous separable convolution for semantic image segmentation, in *Proceedings of the European Conference on Computer Vision (ECCV)*, 2018, pp. 801–818.
6. S. Hochreiter and J. Schmidhuber, Long short-term memory, *Neural Computation*. **9**(8), 1735–1780 (1997).
7. G. Van Houdt, C. Mosquera, and G. Nápoles, A review on the long short-term memory model, *Artificial Intelligence Review*. **53**(8), 5929–5955 (2020).

8. Z. Wang and T. Oates, Imaging time-series to improve classification and imputation, in *Twenty-Fourth International Joint Conference on Artificial Intelligence*, 2015, pp. 3939–3945.
9. L. Lacasa, B. Luque, F. Ballesteros, J. Luque, and J. C. Nuno, From time series to complex networks: The visibility graph, *Proceedings of the National Academy of Sciences.* **105**(13), 4972–4975 (2008).
10. C. Shang, Q. Liu, K.-S. Chen, J. Sun, J. Lu, J. Yi, and J. Bi, Edge attention-based multi-relational graph convolutional networks, (2018).
11. H. Hu, Q. Bai, and Y. Xu, SCSGuard: Deep scam detection for ethereum smart contracts, in *IEEE INFOCOM 2022-IEEE Conference on Computer Communications Workshops (INFOCOM WKSHPS)*, 2022, pp. 1–6.
12. Z. Jin, X. Shi, Y. Yang, X. Yin, Z. Wang, and J. Wu, Toposcope: Recover as relationships from fragmentary observations, in *Proceedings of the ACM Internet Measurement Conference*, 2020, pp. 266–280.
13. A. Addawood, R. Rezapour, O. Abdar, and J. Diesner, Telling apart tweets associated with controversial versus non-controversial topics, in *Proceedings of the Second Workshop on NLP and Computational Social Science*, 2017, pp. 32–41.
14. A.-M. Popescu and M. Pennacchiotti, Detecting controversial events from Twitter, in *Proceedings of the 19th ACM International Conference on Information and Knowledge Management*, 2010, pp. 1873–1876.
15. K. He, X. Zhang, S. Ren, and J. Sun, Delving deep into rectifiers: Surpassing human-level performance on imagenet classification, in *Proceedings of the IEEE International Conference on Computer Vision*, 2015, pp. 1026–1034.
16. J. Deng, W. Dong, R. Socher, L.-J. Li, K. Li, and L. Fei-Fei, Imagenet: A large-scale image database, in *2009 IEEE Conference on Computer Vision and Pattern Recognition*, 2009, pp. 248–255.
17. R. Girshick, Fast R-CNN, in *Proceedings of the IEEE International Conference on Computer Vision*, 2015, pp. 1440–1448.
18. O. Ronneberger, P. Fischer, and T. Brox, U-net: Convolutional networks for biomedical image segmentation, in *International Conference on Medical Image Computing and Computer-Assisted Intervention*, 2015, pp. 234–241.
19. N. E. West and T. O'Shea, Deep architectures for modulation recognition, in *2017 IEEE International Symposium on Dynamic Spectrum Access Networks (DySPAN)*, 2017, pp. 1–6.
20. T. J. O'Shea, J. Corgan, and T. C. Clancy, Convolutional radio modulation recognition networks, in *International Conference on Engineering Applications of Neural Networks*, 2016, pp. 213–226.

21. A. Jagannath, J. Jagannath, and P. S. P. V. Kumar, A comprehensive survey on radio frequency (RF) fingerprinting: Traditional approaches, deep learning, and open challenges, arXiv preprint arXiv:2201.00680 (2022).

22. M. Soltani, V. Pourahmadi, A. Mirzaei, and H. Sheikhzadeh, Deep learning-based channel estimation, *IEEE Communications Letters.* **23** (4), 652–655 (2019).

23. D. Yu and L. Deng, *Automatic Speech Recognition*, 2016, vol. 1, London: Springer.

24. E. H. Houssein, M. Kilany, and A. E. Hassanien, ECG signals classification: A review, *International Journal of Intelligent Engineering Informatics.* **5**(4), 376–396 (2017).

25. Q. Xuan, J. Wang, M. Zhao, J. Yuan, C. Fu, Z. Ruan, and G. Chen, Subgraph networks with application to structural feature space expansion, *IEEE Transactions on Knowledge and Data Engineering.* **33**(6), 2776–2789 (2019).

26. J. Wang, P. Chen, S. Yu, and Q. Xuan, TSGN: Transaction subgraph networks for identifying ethereum phishing accounts, in *International Conference on Blockchain and Trustworthy Systems*, 2021, pp. 187–200.

27. T. E. Ng and H. Zhang, Predicting internet network distance with coordinates-based approaches, in *Proceedings. Twenty-First Annual Joint Conference of the IEEE Computer and Communications Societies*, 2002, vol. 1, pp. 170–179.

28. Q. Xuan, X. Shu, Z. Ruan, J. Wang, C. Fu, and G. Chen, A self-learning information diffusion model for smart social networks, *IEEE Transactions on Network Science and Engineering.* **7**(3), 1466–1480 (2019).

29. J. Shen, J. Zhou, Y. Xie, S. Yu, and Q. Xuan, Identity inference on blockchain using graph neural network, in *International Conference on Blockchain and Trustworthy Systems*, 2021, pp. 3–17.

30. S. Peng, X. Shu, Z. Ruan, Z. Huang, and Q. Xuan, Inferring multiple relationships between ases using graph convolutional network, arXiv preprint arXiv:2107.13504 (2021).

31. S. Dori-Hacohen and J. Allan, Automated controversy detection on the web, in *European Conference on Information Retrieval*, 2015, pp. 423–434.

About the Editors

Qi Xuan received his BS and PhD degrees in control theory and engineering from Zhejiang University, Hangzhou, China, in 2003 and 2008, respectively. He was a Post-Doctoral Researcher with the Department of Information Science and Electronic Engineering, Zhejiang University, from 2008 to 2010, and a Research Assistant with the Department of Electronic Engineering, City University of Hong Kong, Hong Kong, in 2010 and 2017. From 2012 to 2014, he was a Post-Doctoral Fellow with the Department of Computer Science, University of California at Davis, CA, USA. He is a Senior Member of IEEE, and is currently a Professor with the Institute of Cyberspace Security, College of Information Engineering, Zhejiang University of Technology, Hangzhou, China. His current research interests include network science, graph data mining, cyberspace security, machine learning, and computer vision.

Yun Xiang received his BS degree in physics from Zhejiang University, China, in 2006 and a PhD degree in Electrical Engineering from University of Michigan, USA, in 2014. He is currently a Lecturer with the Institute of Cyberspace Security, Zhejiang University of Technology, Hangzhou, China. His current research interests include cyberspace security, machine learning, embedded systems, and computer vision.

Dongwei Xu received his BE and PhD degrees from State Key Laboratory of Rail Traffic Control and Safety, Beijing Jiaotong University, Beijing, China, in 2008 and 2014, respectively. He is currently an Associate Professor with Institute of Cyberspace Security, Zhejiang University of Technology, Hangzhou, China. His research interests include intelligent transportation control, management, and traffic safety engineering.

Contents

Introduction

This book introduces several advanced deep learning algorithms and their applications under various circumstances. The contents are divided into three parts. The first part introduces typical methods in vision applications, such as convolutional neural network (CNN), for the tasks including PM2.5 estimations (Chapter 1), ship license recognition (Chapter 2), defect detection (Chapter 3), and crop stress analysis (Chapter 4). After that, the application in signal processing is described. In particular, in Chapter 5, we give a mixed pruning method for modulation recognition model. Broad learning is introduced for time series data classification (Chapter 6). In Chapter 7, we give a deep learning denoising method for radio modulation signals, and introduce an improved visibility graph for modulation recognition (Chapter 8). Then, we introduce deep learning applications on graph data. In Chapter 9, we demonstrate the Autonomous System (AS) classification with graph convolutional networks (ASGNN), in Chapter 10, we introduce a graph-embedding method for social media opinions analysis, in Chapters 11 and 12 we use GNN to detect Ponzi scheme in ethereum transaction and predict molecular biological activity, respectively.

Part I: Vision Applications

Chapter 1 will introduce a vision-based deep learning model to estimate the PM2.5 concentrations. The inputs of the proposed model are six feature maps extracted from the original image, including refined dark channel, max local contrast, max local saturation, min local color attenuation, hue disparity, and chroma. The model extracts the haze information from the input feature maps and outputs the final PM2.5 concentrations.

Chapter 2 will introduce a ship plate identification technique based on R2CNN and ASTER, which utilizes a two-stage end-to-end network, including text detection, rectification, and recognition. The network is demonstrated on a dataset built from Xiangshan port. It achieves high performance and can identify the ship in real time.

Chapter 3 will introduce two methods to identify the surface defects with different detection granularity. The first one utilizes a deep learning network for dichotomy and uses a generative adversarial network (GAN) for data enhancement. The experimental results show that the deep learning-based method has a high detection efficiency and GAN improves the detection performance of the network. The second method is a fine-grained defect detection network, which mainly focuses on the detection of subtle scratches. This method is divided into two stages. The first stage consists of a Faster R-CNN network and the second stage uses DeepLabV3+ network to detect scratches.

Chapter 4 will explore the research of deep learning in agriculture crop stress analysis. To better illustrate the issue, this chapter begins with the stress types and challenges of identifying stress. Then it introduces the deep neural networks used in agriculture. Finally, it concludes with a summary of the current situation, limitations and future work.

Part II: Signal Applications

Chapter 5 will propose a mixed pruning method, which combines the filter-level and the layer-level pruning method. First, we apply the filter-level and the layer-level pruning method to the signal modulation recognition model. Second, we choose the appropriate model after pruning and further compress the CNN model. The experimental results show that the mixed pruning method has a good performance in the compression of the CNNs.

Chapter 6 will introduce a novel method (namely, GAF–BLS) to classify time series data. In this method, the time features and deep abstract features of the time series data are extracted through the gram angle field, and then in the broad learning system, these features are mapped to a more discriminative space to further enhance them. Experimental results show that the GAF–BLS algorithm can significantly improve the recognition efficiency on the basis that the recognition rate is not reduced.

Chapter 7 will introduce an electromagnetic modulation signal denoising technology based on deep learning, that is, an end-to-end generative model, which makes full use of deep learning to automatically extract features and simplify the signal denoising step. The generative model is based on generative adversarial network, and its loss function adds average error loss and continuity loss in line with signal characteristics on the basis of traditional LSGAN loss. Experiments have proved that this method is better than the low-pass filter in the time and frequency domain, has good generalization ability and can denoise different modulation signals.

Chapter 8 will introduce a new framework based on Graph Neural Network (GNN). First, an improved VG method named Local Limited Penetrable Visibility Graph (LLPVG) is proposed for representing radio signals. Second, a universal multi-channel graph feature extraction GNN model is proposed for corresponding I/Q signal graphs. Finally, we carry out experiments on the radio signal common datasets to prove the powerful representation ability of the proposed methods, and the results show that our modulation recognition frameworks yield more excellent performance compared to typical VG methods and other radio signal classification models.

Part III: Network Applications

Chapter 9 focuses on Autonomous System (AS) business type as a new feature of AS relationship inference. Therefore, the new framework AS-GNN that we have proposed has two main tasks: (1) The static properties of AS and its structural properties in the network are used for AS Classification; (2) The framework takes into account the AS type, global network structure, and local link features concurrently to inference AS relationship. The experiments on real Internet topological data validate the effectiveness of our method. AS-GNN outperforms a series of baselines on both AS Classification and AS Relationship tasks, and can help to further understanding the structure and evolution of the Internet.

Chapter 10 will introduce an end-to-end graph embedding method comment-tree2vec. This method is applied for controversy detection. comment-tree2vec integrates dynamic and static structure features and uses the comment tree structure for improvement. The comment tree is a graph model constructed for characterizing the relationship between the connected entities, which includes comments, users, and news. Besides, the extensive experiments on Toutiao datasets show the effectiveness of comment-tree2vec on controversy detection compared with other methods.

Chapter 11 mainly focuses on the Ponzi scheme, a typical fraud, which has caused large property damage to the users of Ethereum. We model the identification and detection of the Ponzi scheme as a node classification task. This chapter contrasts the differences between Ponzi schemes and normal trading patterns, proposing a Ponzi scheme detection framework based on trading subgraph, graph neural network (GNN), and data augmentation, in which the subgraph mini-batch training can alleviate the limitations of computation requirement and be scalable to large networks. Data augmentation improves the generalization of the model to various Ponzi scheme patterns. The experimental results show that the proposed method is effective.

The drug development cycle is long and costly, and the use of computer methods to screen lead compounds can effectively improve its efficiency. Among them, the Quantitative structure–activity relationship modeling methods to predict the biological activity of molecules is a large area. Improving its prediction accuracy can greatly accelerate the speed of drug development. However, due to the limitation of input methods and computing power, the machine learning modeling methods used in the early days can no longer meet the needs of the drug big data era. The purpose of Chapter 12 is to construct a more reliable prediction model of molecular biological activity. First of all, the instability and unreliability caused by artificial computing features are avoided by learning molecular graph features directly. Secondly, in the process of modeling, it is found that there are some problems, such as the inability to adaptively learning the weight of features in feature fusion and the imbalance of difficult and easy samples in data, thus affecting the overall performance of the model. In this chapter, a graph convolutional network architecture based on edge attention is applied to the biological activity dataset selected to learn molecular graph directly, so as to avoid errors caused by artificial feature engineering. This model shows better classification performance than traditional machine learning methods, and its Accuracy index can be 2–8% higher.

Part I
Vision Applications

Chapter 1

Vision-Based Particulate Matter Estimation

Kaihua Zhang, Zuohui Chen, and Yun Xiang*

Institute of Cyberspace Security,
Zhejiang University of Technology,
Hangzhou, P.R. China

*xiangyun@zjut.edu.cn

The monitoring of air quality index (AQI), especially particulate matter 2.5 (PM2.5), in urban areas has drawn considerable attention due to the increasing air pollution. However, existing PM2.5 measurement relies on air monitoring stations, which are expensive and distributed sparsely in the city. Areas far away from the monitoring stations can hardly get accurate PM2.5 readings. Hence, we need a fine-grained estimation of PM2.5. We propose a vision-based PM2.5 estimation method. It leverages multi-view images and haze-related features to train a neural network. We also collect a dataset with images and high-resolution PM2.5 data. Experimental results show that our estimated mean square error is only 2.27 μgm^{-3}.

1. Introduction

Air pollution is a serious threat to human health. It is closely related to disease and death rates.[1-5] The World Health Organization

3

estimates that 2.4 million people die annually from causes associated with air pollution.[6] The most common air pollutants are particulate matter (PM), sulfur dioxide, and nitrogen dioxide. This work focuses on PM2.5, which can increase the rate of cardiovascular, respiratory, and cerebrovascular diseases.[7,8] Air monitoring stations are now used to estimate PM2.5, which are correlated with pollutant concentrations.[9] However, the limited number of sensors and, therefore, low spatial density makes them inaccurate.

Low measurement spatial density makes it especially difficult to estimate human exposures. PM has heterogeneous sources,[10] e.g., automobiles exhaust, dust, cooking, manufacturing, and building construction, etc. The PM concentrations are correlated with source distributions. For example, numerous factors, including wind, humidity, and geography,[11,12] are related to PM distributions. Therefore, air pollution concentration varies within a relatively short distance: relying on existing sparse, stationary monitoring stations can lead to inaccurate estimation of the high-resolution pollution field.

Increasing sensor density or adding image sensors supporting high spatial resolution captures can increase estimation accuracy and resolution. PM2.5 can be estimated by analyzing the visual haze effect caused by particles and gasses.[13] The image data may be derived from several sources such as social media[14] and digital cameras.[15] The ground truth data are typically derived from the nearest air quality station and have low spatial resolution. Existing approaches are generally inaccurate except near the sparsely deployed sensing stations. Moreover, they generally assume homogeneous distributions of particles and gases within images, implying consistent light attenuation. However, in reality, pollution concentration varies rapidly in space. Thus, accurate evaluation requires vision-based estimation algorithms.

In this chapter, we present a vision-based PM2.5 estimation algorithm and collect an air pollution dataset containing images to evaluate our algorithm. The algorithm consists of two steps, haze feature extraction and PM concentration estimation, where haze feature is represented by six haze-relevant image features and PM concentration is estimated by a deep neural network. Our dataset contains images captured by a drone and ground PM concentration measured by particle sensors. The main contents of this chapter are summarized as follows:

(1) We propose a deep neural network to estimate PM2.5 concentration. Its input combines both vision and ground sensor information.

(2) We collect an air pollution dataset containing both high spatial resolution PM measurement and corresponding images.

(3) The proposed algorithm is evaluated on our dataset, experimental result shows that it achieves remarkable performance ($2.27\ \mu\text{gm}^{-3}$ by root mean-squared error (RMSE)).

2. Related Work

2.1. *Vision-based air quality estimation*

Nowadays image data are easy to access, which is greatly convenient for the vision-based air pollution estimation algorithms. Zhang et al.[16] propose an ensemble CNN for vision-based air quality estimation. They collect their own images and official air quality data to train the models. The input of each classifier is the original images and the output is the air quality index (AQI). By dividing air quality into different levels, they turn the estimation problem into a classification problem. Instead of estimating the air quality, our work estimates the air quality concentrations directly. Li et al.[14] use social media photos to monitor the air quality. They combine transmission and depth of pixels in the image to estimate the haze level. The extracted features are manually designed. Yang et al.[17] propose Imgsensingnet, which is a CNN for predicting the AQI. It takes six image feature maps as input and air quality level as output. They use a drone and a ground sensor network to take pictures and ground truth data for prediction. The main difference between our work and theirs is the network structure, network output, and input data. We use 2D convolution as the basic component and output PM concentration instead of AQI. Meanwhile, in addition to the feature maps, the model input also includes sensor readings.

2.2. *Air quality dataset*

Air pollution has a high-spatial variation.[18,19] However, existing air quality datasets do not contain both high-resolution air quality concentration and corresponding images. Li et al.[20] collect a database including images in different outdoor scenes, captured regularly for 5 years. There are a total of 7,649 images from 85 different view sites in 26 cities in China. For air pollutant, their data only provide the AQI instead of the raw pollutant concentration. Apte *et al.*[21] use Google street view vehicles to sample air quality data in a 30 km area of Oakland, CA, which contain daytime NO, NO_2, and black carbon. Though the dataset is of high-resolution, corresponding images of streets and PM are not included. Zhang *et al.*[22] propose Airnet for air quality forecasting based on machine learning. The dataset contains multiple meteorological data and pollutants. However, their air quality data comes from monitoring station, which means it also has low spatial resolution.

Therefore, we propose to collect a dataset satisfying the need of images and high-resolution PM2.5 data simultaneously.

3. Methodology

We propose PMEstimatingNet, which performs video-based estimating utilizing the pictures obtained by a UAV. To quantify PM2.5 values, we introduce haze-relevant features for raw haze images. Then based on the extracted haze features, a novel CNN model is designed for further processing feature maps and predicting PM2.5 values.

3.1. *Haze image processing*

In computer vision, Eq. (1) is widely used to describe the formation of a hazy image.

$$\mathbf{I}(x) = \mathbf{J}(x)t(x) + L_\infty(1 - t(x)), \tag{1}$$

where $\mathbf{I}$ is the observed hazy image, $\mathbf{J}$ is the image after eliminating haze, t represents the medium transmission, L_∞ is the global atmospheric light, and x denotes pixel coordinates. In contrast to haze-removal methods, the feature maps-based methods provide another approach to estimating the value of haze in a single image. The information related to the content is removed and the information related to the haze concentration is retained in the feature maps. In other words, our approach concentrates on the discrepancy between $\mathbf{I}(x)$ and $\mathbf{J}(x)$.

3.2. Haze-relevant features extraction

First, we extract different kinds of haze-relevant statistical features. We list six content-unrestricted and haze-relevant features[17] as illustrated in Fig. 1.

(1) *Refined dark channel*: Dark channel,[23] defined as the minimum of all pixel colors in a local patch, is an informative feature for haze detection.

$$D(x, \mathbf{I}) = \min_{\mathbf{y} \in \Omega(x)} \left(\min_{c \in \{r,g,b\}} \frac{\mathbf{I}^c(y)}{L_\infty^c} \right), \tag{2}$$

where $\Omega(x)$ is a local patch centered at x and I^c is one color channel of $\mathbf{I}$. Most local patches in outdoor haze-free images contain

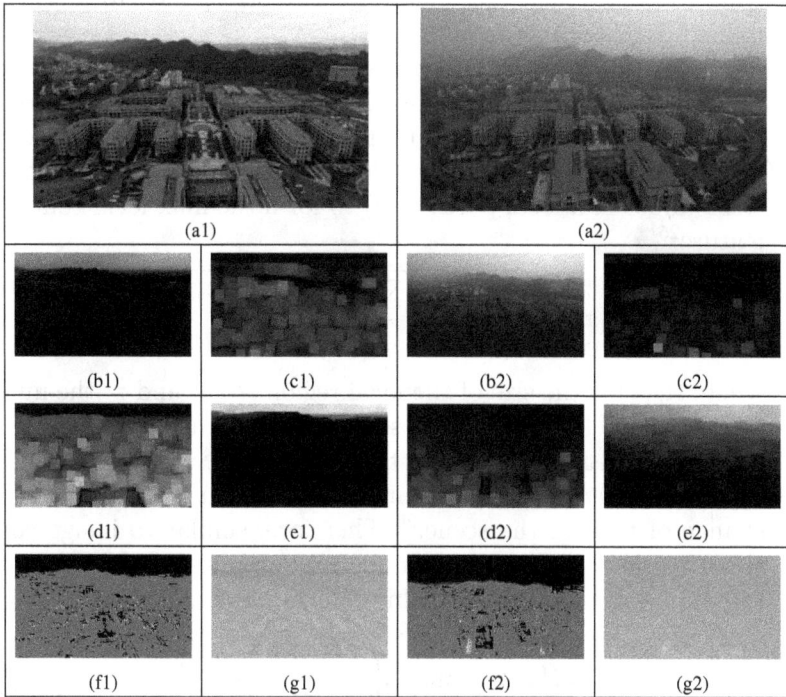

Fig. 1: An example of the extracted feature maps for two raw haze images (the PM value of image 1 is 6 μgm^{-3}, the PM value of image 2 is 118 μgm^{-3}): (a) origin input image; (b) refined dark channel; (c) max local contrast; (d) max local saturation; (e) min local color attenuation; (f) hue disparity; (g) chroma.

some pixels with intensity very low in at least one color channel.[23] Therefore, the dark channel can roughly reflect the thickness of the haze.

For better estimation of haze value, refined dark channel[17] is proposed with the application of filter G on the estimated medium transmission $\tilde{t}$ [24] to identify the sharp discontinuous edge and draw the haze profile. Note that by applying the minimum operation on Eq. (1), the dark channel of J tends to be zero, i.e., $\tilde{t} = 1 - D(x; I)$.

The refined dark channel is

$$D^{\mathrm{R}}(x; \mathbf{I}) = 1 - G \left(1 - \min_{y \in \Omega(x)} \left(\min_{c \in \{r,g,b\}} \frac{\mathbf{I}^C(y)}{L_\infty^c} \right) \right), \qquad (3)$$

Figure 1(b) shows the refined dark channel feature. As illustrated in the image, the hazer image corresponds to a whiter refined dark channel map.

(2) *Max local contrast*: Due to the scattering effect, the contrast of the image can be reduced by the haze significantly. The local contrast is defined as the variance of pixel intensities in a local $r \times r$ square. We further use the local maximum of local contrast values in a local patch $\Omega(x)$ to form the max local contrast feature[25]

$$C^{\mathrm{T}}(x; \mathbf{I}) = \max_{y \in \Omega(x)} \sqrt{\frac{\sum_{z \in \Omega_r(y)} \|I(z) - I(y)\|^2}{\pi |\Omega_r(y)|}}, \qquad (4)$$

where $\Omega_r(y)$ is the size of the local region $\Omega_r(y)$ and π, the number of channels. Figure 1(c) shows the contrast feature of the correlation between haze and the contrast feature.

(3) *Max local saturation*: The image saturation varies with the change of haze in the scene.[26] Therefore, similar to image contrast, the max local saturation feature[17] represents the maximal saturation value of pixels within a local patch.

$$S(x; \mathbf{I}) = \max_{y \in \Omega(x)} \left(1 - \frac{\min_c \mathbf{I}^c(y)}{\max_c \mathbf{I}^c(y)} \right). \qquad (5)$$

The max local saturation feature for the mountains, buildings, and meadows in the images are significantly different as shown in Figs. 1(d1) and 1(d2). It is also correlated to the haze density.

(4) *Min local color attenuation*: The scene depth is positively correlated with the difference between the image brightness and saturation.[27] The scene depth is represented by the color attenuation prior.

$$d\left(x;\mathbf{I}\right) = \theta_0 + \theta_1 \cdot I^{\mathrm{v}}(x;\mathbf{I}) + \theta_2 \cdot I^{\mathrm{s}}(x;\mathbf{I} + \epsilon(x;\mathbf{I}), \qquad (6)$$

where I^{v} and I^{s} are the brightness and the saturation, respectively. Assuming $\epsilon(x) \sim \mathcal{N}(0,\sigma^2)$, θ_0, θ_1, θ_2, and σ can be estimated through maximum likelihood method. The min local color attenuation feature[17] can be obtained for better representation of the haze influence from the raw depth map considering the minimum pixel-wise depth within a local patch $\Omega(x)$.

$$A(\mathbf{x};\mathbf{I}) = \min_{y \in \Omega(x)} d(y;\mathbf{I}). \qquad (7)$$

Figure 1(e) shows the min local color attenuation feature, which presents a visual correlation with haze density.

(5) *Hue disparity*: In Ref. 28, authors utilize the hue disparity between the original image and its semi-inverse image for haze removal. The semi-inverse image is defined as the max value between original image and its inverse.

$$I_{\mathrm{si}}^{\mathrm{c}}(x;\mathbf{I}) = \max_{x \in I, c \in \{r,g,b\}} \{I_{\mathrm{c}}(x), 1 - I_{\mathrm{c}}(x)\}. \qquad (8)$$

The hue disparity is also reduced under the influence of haze. Thus, it can be used as another haze-relevant feature.

$$H\left(x;I\right) = |I_{\mathrm{si}}^{\mathrm{h}}(x;I) - I^{\mathrm{h}}(x;I)|, \qquad (9)$$

where I^{h} is the hue channel of the image. Figure 1(f) shows the hue disparity feature for image with haze.

(6) *Chroma*: In the CIELab color space, one of the most representative features to describe the color degradation in the atmosphere is the chroma. Assume $[L(x;I), a(x;I), b(x;I)]^{\mathrm{T}}$ are the haze image I in the CIELab space, the chroma feature is defined as

$$C^{\mathrm{H}}(x;\mathbf{I}) = \sqrt{a^2(x;\mathbf{I}) + b^2(x;\mathbf{I})}. \qquad (10)$$

As shown in Fig. 1(g), chroma is correlated to the haze density.

3.3. *Data processing*

Single image contains noise and has a limited visual field. Using images of different views can mitigate the influence of noise and acquire more information.

To take advantage of images from multiple views, we divide the whole video into six parts equally. The input of our CNN contains six images from each part. For all the images, we resize them to 108×192 with bilinear interpolation. Then we generate haze-related features for all the six images. 36 haze-related features are obtained from the original image data as the input of our network. Figure 1 shows the extracted feature maps. We observe significant differences between hazed images and clean images.

3.4. *Model architecture*

CNN is widely utilized in image processing and visual applications. Due to its ability of extracting complex features, it has achieved great successes in many fields. We design a video-based CNN network to predict PM2.5 value from multiple feature maps.

As shown in Fig. 2, the extracted feature has a dimension of $108 \times 192 \times 36$. For the first layer we set kernel size to 3×3. In the subsequent pooling layer, 2×2 max pooling is applied. Our model consists of three Conv-pooling blocks. The flatten vector is

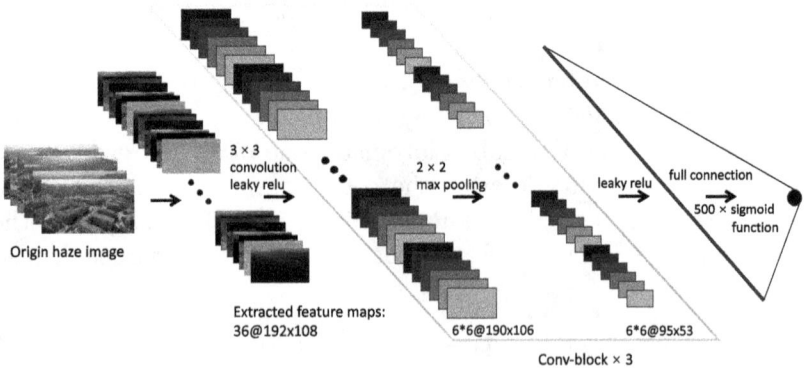

Fig. 2:   The architecture of the proposed CNN model.

fully connected to the output layer. We choose the leaky ReLU function as the activation function, which helps to avoid the dying ReLU phenomenon.

4. Experiment

4.1. Devices

(1) *PM device*: Nova PM sensor modules and Jetson nano are combined for PM2.5 collection. The equipment collects the PM2.5 value at an interval of several seconds.
(2) *UAV*: DJI Air 2S is utilized for videos capture: [https://www.dji.com/air-2s/specs].

The specific parameters of PM Device and UAV are shown in Table 1.

Table 1:   Parameters of PM device and UAV.

Nova PM sensor	
Sensor range	[PM2.5] 0.0–999.9 $\mu\mathrm{gm}^{-3}$
	[PM10] 0–1999.9 $\mu\mathrm{gm}^{-3}$
Operating temperature	-10–$50°$C
Operating humidity	Maximum 70%
The response time	1 s
Serial port data output frequency	1 Hz
Minimum resolution particle size	0.3 μm
The relative error	Max. $\pm15\%$ and ±10 $\mu\mathrm{gm}^{-3}$
	(*Note*: 25°C, 50%RH)
Standard certification	CE/FCC/RoHS
UAV camera	
Sensor	1$''$ CMOS
	Effective pixels: 20 MP; 2.4 μm pixel size
Lens	FOV: 88°
	35 mm format equivalent: 2 mm
	Aperture: $f/2.8$
	Shooting range: 0.6 m to ∞
Video resolution	MP4 (H.264/MPEG-4 AVC, H.265/HEVC)
The lens angle of depression	17°~19°

4.2.   Experiment setup and data collection

(1) *Layout of PM2.5 sensors*: In order to obtain fine-grained PM2.5
    values in the monitoring area, nine PM sensors are placed in
    the area as shown in Fig. 3(b). We use the average value dur-
    ing the experiment as the ground truth value. The PM2.5 value
    ranges from 6 to 118. The distribution of sensors is shown in the
    Fig. 3(a).
(2) *Image Data*: We use a UAV (Fig. 4) to obtain 12 videos. The
    length of each video is over 3,000 s with resolution of 1080×1920.

4.3.   Evaluation metrics and inference performance

The RMSE is used as the evaluation criteria.

$$\text{RMSE} = \sqrt{\frac{1}{N}\sum_{i=1}^{N}(y_i^p - y_i^t)^2}, \tag{11}$$

where N is the total number of samples, y_i^p is the predicted value of
the ith sample, and y_i^t is the ground truth of the ith sample.

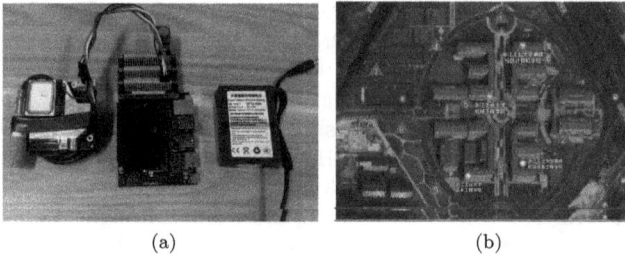

(a)   (b)

Fig. 3:   One of PM sensors (a) and their layout (b).

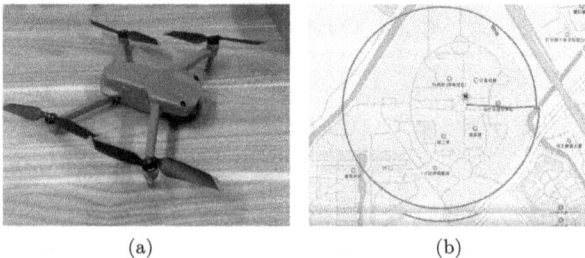

(a)   (b)

Fig. 4:   UAV (a) and its flight trajectory (b).

Figure 5(a) shows the prediction error during training. As the number of iterations increases, the error gradually decreases. When the number of iterations is greater than 15, the difference between the training error and the test error is small. Thus, we stop training after 18 iterations.

In order to evaluate our method, we perform a 12-fold cross-validation, where one video is for testing while others are for training. The mean of multiple verification results is 2.27 μgm^{-3}.

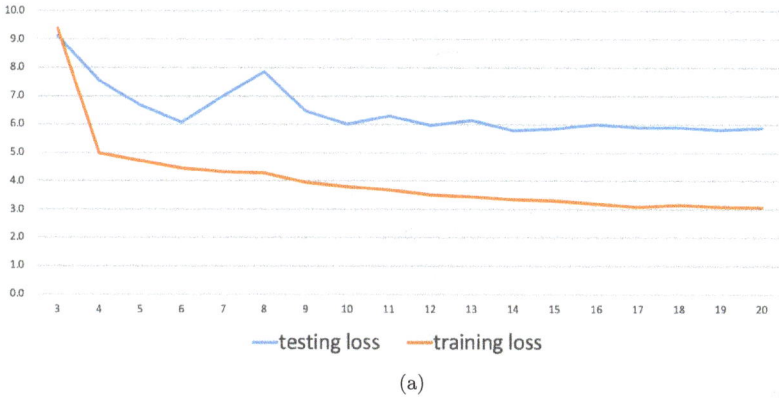

(a)

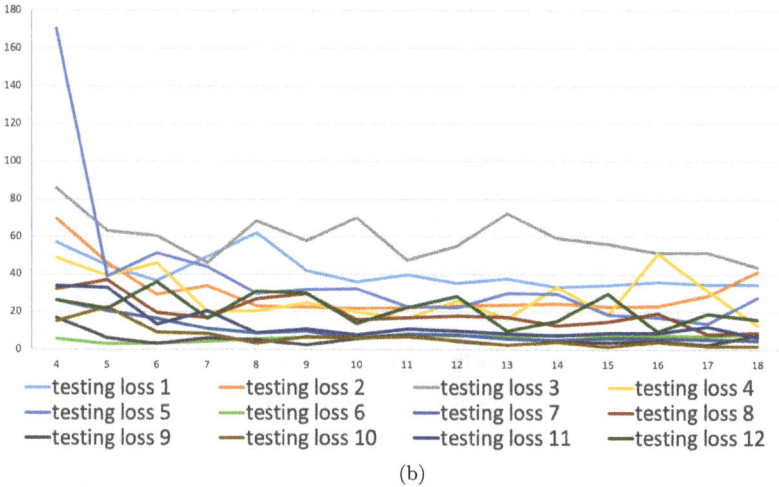

(b)

Fig. 5:   The horizontal axis represents the number of iterations, and the vertical axis represents RMSE (a) training error and test error under different iterations under one training; (b) test error of different iteration times under 12-fold cross-validation.

We verify the effect of multi-view images using pictures at the same location as the input of our model. The training is set to the same configuration. The average value of the multiple verification results is 3.89 μgm^{-3}, as shown in Fig. 6.

(a)

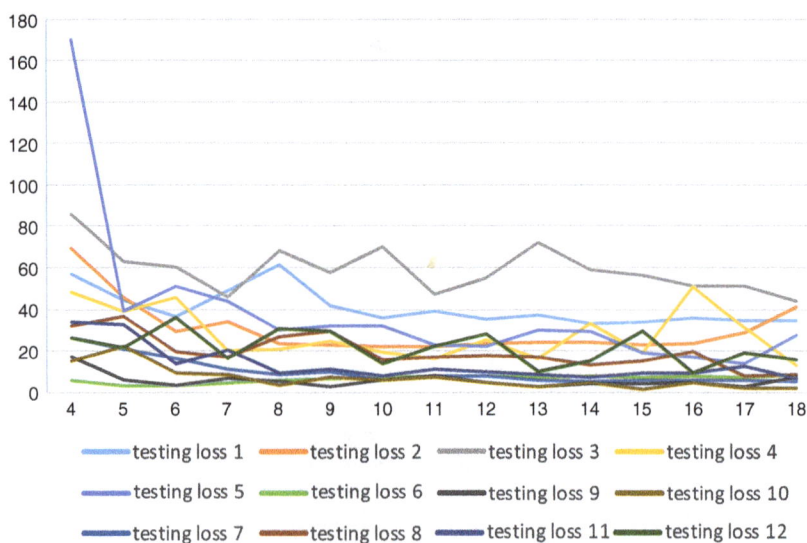

(b)

Fig. 6: The horizontal axis represents the number of iterations, and the vertical axis represents RMSE (a) training error and test error under different iterations under one training; (b) test error of different iteration times under 12-fold cross-validation.

5. Conclusion

We propose a vision-based PM2.5 estimating method for fine-grained air monitoring. Our method utilizes multi-view images and haze-relevant features. We collect a dataset containing both images and high-resolution PM2.5 data for training and validation. Experimental results demonstrate that our method is effective and achieved 2.27 μgm^{-3} RMSE on our dataset.

References

1. K. Naddafi, M. S. Hassanvand, M. Yunesian, F. Momeniha, R. Nabizadeh, S. Faridi, and A. Gholampour, Health impact assessment of air pollution in megacity of Tehran, Iran, *Iranian Journal of Environmental Health Science & Engineering.* **9**(1), 28 (2012).
2. H. Nourmoradi, Y. O. Khaniabadi, G. Goudarzi, S. M. Daryanoosh, M. Khoshgoftar, F. Omidi, and H. Armin, Air quality and health risks associated with exposure to particulate matter: A cross-sectional study in Khorramabad, Iran, *Health Scope.* **5**(2), e31766 (2016).
3. S. Poduri, A. Nimkar, and G. S. Sukhatme, Visibility monitoring using mobile phones, *Annual Report: Center for Embedded Networked Sensing.* 125–127 (2010).
4. B. Chen and H. Kan, Air pollution and population health: A global challenge, *Environmental Health and Preventive Medicine.* **13**(2), 94–101 (2008).
5. F. J. Kelly and J. C. Fussell, Air pollution and public health: Emerging hazards and improved understanding of risk, *Environmental Geochemistry and Health.* **37**(4), 631–649 (2015).
6. I. Abubakar, T. Tillmann, and A. Banerjee, Global, regional, and national age-sex specific all-cause and cause-specific mortality for 240 causes of death, 1990-2013: A systematic analysis for the Global Burden of Disease Study 2013, *The Lancet.* **385**(9963), 117–171 (2015).
7. S. Feng, D. Gao, F. Liao, F. Zhou, and X. Wang, The health effects of ambient PM2.5 and potential mechanisms, *Ecotoxicology and Environmental Safety.* **128**, 67–74 (2016).
8. J. O. Anderson, J. G. Thundiyil, and A. Stolbach, Clearing the air: A review of the effects of particulate matter air pollution on human health, *Journal of Medical Toxicology.* **8**(2), 166–175 (2012).
9. R. A. Rohde and R. A. Muller, Air pollution in China: Mapping of concentrations and sources, *PLoS one.* **10**(8), 1–14 (2015).

10. Y. Zhang, J. Cai, S. Wang, K. He, and M. Zheng, Review of receptor-based source apportionment research of fine particulate matter and its challenges in China, *Science of the Total Environment.* **586**, 917–929 (2017).
11. G. Lin, J. Fu, D. Jiang, W. Hu, D. Dong, Y. Huang, and M. Zhao, Spatio-temporal variation of PM2.5 concentrations and their relationship with geographic and socioeconomic factors in China, *International Journal of Environmental Research and Public Health.* **11**(1), 173–186 (2014).
12. X. Zhao, W. Zhou, L. Han, and D. Locke, Spatiotemporal variation in PM2.5 concentrations and their relationship with socioeconomic factors in China's major cities, *Environment International.* **133**, 105145 (2019).
13. T. Zhang and R. P. Dick, Estimation of multiple atmospheric pollutants through image analysis, in *Proceedings of International Conference on Image Processing*, 2019, pp. 2060–2064.
14. Y. Li, J. Huang, and J. Luo, Using user generated online photos to estimate and monitor air pollution in major cities, in *Proceedings of International Conference on Internet Multimedia Computing and Service*, 2015, pp. 1–5.
15. C. Liu, F. Tsow, Y. Zou, and N. Tao, Particle pollution estimation based on image analysis, *PLoS One.* **11**(2), e0145955 (2016).
16. C. Zhang, J. Yan, C. Li, H. Wu, and R. Bie, End-to-end learning for image-based air quality level estimation, *Machine Vision and Applications.* **29**(4), 601–615 (2018).
17. Y. Yang, Z. Hu, K. Bian, and L. Song. Imgsensingnet: UAV vision guided aerial-ground air quality sensing system, in *IEEE INFOCOM 2019 — IEEE Conference on Computer Communications*, 2019, pp. 1207–1215.
18. X. Zhang, E. Craft, and K. Zhang, Characterizing spatial variability of air pollution from vehicle traffic around the Houston Ship Channel area, *Atmospheric Environment.* **161**, 167–175 (2017).
19. H. Z. Li, P. Gu, Q. Ye, N. Zimmerman, E. S. Robinson, R. Subramanian, J. S. Apte, A. L. Robinson, and A. A. Presto, Spatially dense air pollutant sampling: Implications of spatial variability on the representativeness of stationary air pollutant monitors, *Atmospheric Environment: X.* **2**, 100012 (2019).
20. Q. Li and B. Xie. Image-based air quality estimation, in *Chinese Conference on Pattern Recognition and Computer Vision (PRCV)*, 2019, pp. 161–171.

21. J. S. Apte, K. P. Messier, S. Gani, M. Brauer, T. W. Kirchstetter, M. M. Lunden, J. D. Marshall, C. J. Portier, R. C. Vermeulen, and S. P. Hamburg, High-resolution air pollution mapping with Google street view cars: Exploiting big data, *Environmental Science & Technology.* **51**(12), 6999–7008 (2017).
22. S. Zhao, X. Yuan, D. Xiao, J. Zhang, and Z. Li, Airnet: A machine learning dataset for air quality forecasting (2018). http://airnet.caiyunapp.com/.
23. K. He, J. Sun, and X. Tang, Single image haze removal using dark channel prior, *IEEE Transactions on Pattern Analysis and Machine Intelligence.* **33**(12), 2341–2353 (2010).
24. K. He, J. Sun, and X. Tang, Guided image filtering, *Transactions on Pattern Analysis and Machine Intelligence.* **35**(6), 1397–1409 (2012).
25. R. T. Tan, Visibility in bad weather from a single image, in *2008 IEEE Conference on Computer Vision and Pattern Recognition*, 2008, pp. 1–8.
26. S. Li, T. Xi, Y. Tian, and W. Wang, Inferring fine-grained PM2.5 with Bayesian based kernel method for crowdsourcing system, in *GLOBECOM 2017-2017 IEEE Global Communications Conference*, 2017, pp. 1–6.
27. Q. Zhu, J. Mai, and L. Shao, A fast single image haze removal algorithm using color attenuation prior, *IEEE Transactions on Image Processing.* **24**(11), 3522–3533 (2015).
28. C. O. Ancuti, C. Ancuti, C. Hermans, and P. Bekaert, A fast semi-inverse approach to detect and remove the haze from a single image, in *Asian Conference on Computer Vision*, 2010, 501–514.

© 2023 World Scientific Publishing Company
https://doi.org/10.1142/9789811266911_0002

Chapter 2

Automatic Ship Plate Recognition Using Deep Learning Techniques

Hang Xu, Xinhui Zhang, and Yun Xiang*

Institute of Cyberspace Security,
Zhejiang University of Technology,
Hangzhou, P.R. China

**xiangyun@zjut.edu.cn*

Vision-based vehicle identification technology has been widely used in many areas. However, the existing technologies mainly focus on car identification and cannot address the ship identification problem directly due to highly irregular characters and the complicated natural environment. In this chapter, a ship plate identification technique based on R2CNN and ASTER is proposed, which utilizes two-stage end-to-end network, including text detection, rectification, and recognition. The network is demonstrated on a dataset built from Xiangshan port. It achieves high performance and can identify the ship in real time.

1. Introduction

For modern waterway regulation, accurate and timely identification of ships are important. Currently, it is done mostly based on automatic identification system (AIS)[1] using radio technology. The AIS system utilizes the global positioning system (GPS) to broadcast ship position, speed, and course to the shore stations or nearby ships. However, ships without AIS devices and invading ships cannot be identified by the AIS system. In the AIS system, Chinese characters are represented in Chinese pinyin, whose length often exceeds the length limitation of the ship name. Recently, cameras are becoming popular in port management, providing architecture support for vision-based ship identification. Thus, in this work, a deep learning based ship identification method is proposed.

Vision-based vehicle identification technology is already widely used in many areas, such as community vehicle management[2] and traffic intersection violation detection,[3] etc. However, the existing technologies cannot be used to address the ship identification problem directly. The nautical environment is much more complicated. Moreover, the characters of the ship license plate are highly irregular. For example, it is common to have inclined license plates, diverse characters, complex backgrounds, and cover-ups, etc.

Deep learning techniques are powerful and general to address those problems. Compared with the traditional recognition methods, the detection system based on deep learning can typically achieve better performance in complex environments. Researchers have proposed various target detection methods based on deep neural networks. However, ship license plates identification is more complex and mostly unaddressed.

In this chapter, a ship plate identification technique based on R2CNN and ASTER is proposed. The main contributions of this work are as follows.

(1) We collect and build a real-world dataset to train and evaluate our algorithms.
(2) We propose a two-stage end-to-end network which includes text detection, rectification, and recognition.
(3) We deploy and evaluate our network in real-world environments.

The experimental results demonstrate that our system can effectively identify the ship in real time. Our method can achieve precision of 93.4% and 95.9% in detection and recognition stages, respectively.

2. Related Work

2.1. *General object detection*

Traditional target detection algorithms include Hog,[4] DPM,[5] and Viola Jones detector,[6] etc. However, with the emergence of deep convolutional neural network (CNN), a large number of object detection algorithms based on CNN have been proposed.

Target detection algorithms based on deep learning can be divided into two categories, which are one-stage and two-stage, respectively. SSD,[7] DSSD,[8] YOLO,[9] and YOLOv2[10] are all one-stage detection algorithms. One-stage methods do not need region proposal stage. SSD[7] can directly predict the bounding box and get classification of the detected object. In the detection stage, SSD algorithm uses multi-scale method. It generates multiple detection boxes which are close to the ratio of the length and width of the target object on the feature maps. These boxes are used for regression and classification. DSSD[8] is an improved algorithm of SSD. It uses ResNet-101 instead of VGG16 as a feature extraction network. At the same time, it adds the deconvolutional module and the prediction module. YOLO[9] network transforms the target detection problem into a regression one. It divides the image into $s*s$ grid area, and classifies the image of every grid. The training and detection speed of the whole network is fast. YOLOv2[10] applies anchors on the basis of YOLO, and adds a new network structure named Darknet, which improves the performance while maintaining the real-time speed.

Compared with one-stage detection algorithms, two-stage algorithms have better performance but are slower. In the first stage, preliminary detection is performed and regional proposals are obtained. In the later stage, the algorithm executes further detection. There are several mainstream regional proposal methods, including R-CNN,[11] SPP-Net,[12] Fast R-CNN,[13] and Faster R-CNN,[14] etc. R-CNN[11] segments the image from bottom to top and then merges the segmented regions with different scales. Each generated region is a classification candidate. Fast R-CNN uses CNN network to extract the features of the whole image first. Then it creates candidate regions on the extracted feature map. However, the region proposal stage of Fast R-CNN becomes the computing bottleneck of the whole detection network. Thus, Faster R-CNN proposes a special region proposal network (RPN) instead of selective search method, which greatly improves the speed.

2.2. *Text detection*

The characteristics of text make its detection unique from general object detection. First, the length and width ratio of text varies. Second, the text direction must be considered in text detection. Third, texts do not have the obvious closed edge contour like ordinary objects. unsuccessful detection may take one word or part of the text as the result rather than the whole line of text.

East[15] uses full convolutional network (FCN) to generate multi-scale fusion feature map, and then directly performs pixel-level text block prediction. The structure of CTPN[16] is similar to Faster R-CNN. It adds LSTM layer, which can effectively detect the horizontal distribution of text in a complex scene. However, it is not good at non-horizontal text detection. DMPNet[17] uses quads (not rectangles) to label text boxes to more compactly enclose the text area. SegLink[18] first locates part of the text, then merges all parts to form a complete one. It adds rotation angle learning on the basis of SSD. It is difficult to detect text lines with large spacing and curves. Textboxes[19] is an end-to-end text detection model based on SSD framework. Compared with SSD, it adjusts the length–width ratio of the candidate box and the shape of convolutional kernel to a rectangle. Therefore, it is more suitable to detect the narrow text lines. RRPN[20] proposes an anchor with angle based on Faster R-CNN. In the model, the candidate box of text area can be rotated, and the inclination angle of the text line is estimated in the process of border regression calculation. R2CNN[21] is based on the framework of Faster R-CNN. Compared with Faster R-CNN, it adds two pooling size (3×11 and 11×3) in the stage of ROIPooling as well as a regression of inclined box to detect inclined text.

3. Dataset

3.1. *Data collecting*

We construct a dataset of ship license plates called ZJUTSHIP-3656. It is collected from Xiangshan port, Ningbo, China. We sample ship images at various day times and weather conditions. We ensure the diversity of backgrounds, camera angles, and ship types, as shown in Figs. 1(a–c). It can improve the robustness of the algorithms. More details can been found in Table 1.

3.2. *Data labeling*

The labeling procedure is mainly divided into two parts, which are ship plate detection and text recognition, respectively. The labeling of the ship license plate image for detection model training is mainly to mark and record the position of the ship plate. We use the inclined box labeling method due to the inclination of cameras and the ship license plate. We use a polygonal box to mark the four points of the ship license in the image, (x_1, y_1), (x_2, y_2), (x_3, y_3), and (x_4, y_4), which surround the entire ship license plate. We do not discard the ship plates containing incompleteness, occlusion, and blurred characters.

The labeling of text includes character type, character number, and character relationship, etc. Due to the partial occlusion or blurring of the ship license, we leverage the fuzzy labeling to label the image. As shown in Fig. 1(d), we label this ship license image as

Table 1:   Details of datasets.

Camera type	Shooting locations	Shooting duration	Image size	Image format
Hikvision iDS-2DYH4FF	Five different locations at Xiangshan port in Ningbo, China	From 2018 to 2020	1920 × 1080	JPEG

(a)

(b)

(c)

(d)

Fig. 1: Some samples from ZJUTSHIP-3656. (a) Ship images shooting by different cameras. (b) Ship images shooting by same cameras from different angles. (c) Ship images at different times. (d) Case of missing characters in ship license.

XXX45051. In that case, we only label the characters that can be recognized by humans and label the incomplete or severely occluded characters as X.

4. Method

4.1. *Data preprocessing*

Before been processed by the detection algorithms, the RGB images are transformed to grayscale images. This processing can enhance the accuracy at night time. However, it may decrease the recognition accuracy at daytime. Therefore, it is recommended when the network is used at night time.

4.2. *Ship license plate detection*

The purpose of ship license plate detection is to detect the position of ship plate. Generally, the ship license plate is composed of numbers and Chinese characters. Accurate ship plate detection is the precondition of ship plate character recognition. The detection of ship plates could be attributed to the task of text detection in natural scenes. As mentioned in the foregoing section, there are difficulties such as angles, fonts, and background.

R2CNN[21] is a text detection network which is good at detecting inclined text. Our ship license plate detection network is based on it. As shown in Fig. 2, R2CNN framework consists of two modules. One is the region proposal network (RPN), and the other is Fast R-CNN network. RPN is a network that is specially designed to generate region proposals[14] and achieves faster speed compared with the previous methods such as selective search.[22]

The second stage is Fast R-CNN network. Fast R-CNN first obtains the feature map through convolutional layers. In our framework, we deploy ResNet-101[23] as the feature extractor. It converts an input image into a feature map as output. After that, the feature map is used for the RPN structure to obtain the region proposal boxes. Then, the region proposal boxes are maxpooled. At the last step, the loss of classification is output by calculating the category of each proposal through the fully connected layer and softmax layer. At the same time, the position offset of the bounding box is obtained

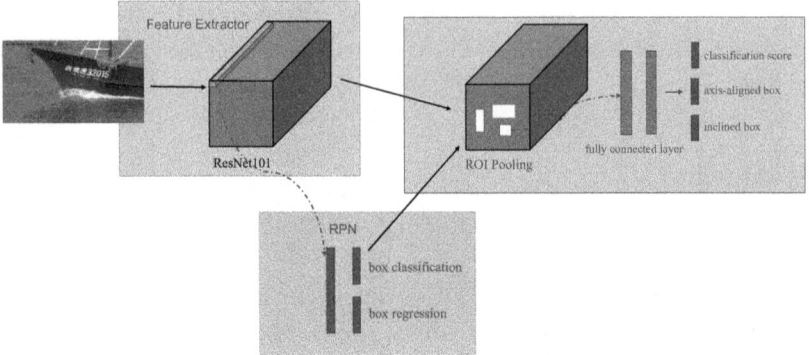

Fig. 2:   R2CNN-based detection network.

by using the bounding box regressor, to regress the target detection box more accurately.

The loss of Fast R-CNN module contains three parts, which are as follows: the classification loss of the two categories, the regression loss of the horizontal box, and the regression loss of the inclined bounding box. The loss function is

$$L = L_{cls} + \lambda_1 \cdot L_{loc_h} + \lambda_2 \cdot L_{loc_r}, \tag{1}$$

where L_{cls} is the classification loss, L_{loc_h} and L_{loc_r} are the regression loss of the horizontal box and the rotational box, respectively.

We optimize parameters of modules for our task. In region proposal stage in RPN module, we set 7 different anchor scales (1, 2, 4, 8, 16, 32, 64) and 13 different anchor ratios (1, 1/2, 2, 1/3, 3, 1/4, 4, 5, 1/5, 6, 1/6, 7, and 1/7), respectively. In the Fast R-CNN network, the size of ROI pooling we use is 7×7.

4.3.   *Text recognition*

In the previous stage, we obtain the candidate box surrounding the ship license plate. In this stage, we apply the ASTER[24] framework to rectify the text of the ship name and recognize the text to get the whole license name. In this section, we introduce our recognition method in detail.

Our recognition network is based on ASTER which is a text recognition framework for irregular texts. It consists of two modules.

The first one is a rectification network, while the second one is a recognition network.

Figure 3 describes the architecture of the rectification network. The rectification network first resizes the image to a proper shape. The control points are generated by a CNN in the next step. Then, a TPS[25] transformation is passed to the grid generator and the sampler. At last, the network generates the rectified image by integrating the origin image and the output of the grid generator.

As shown in Fig. 4, the recognition model consists of an encoder and a decoder. It proposes a bidirectional decoder to enhance the performance. The feature and order of all characters are sensed by the encoder. The encoder first uses convolutional layers to extract a feature map from the image. Then, the network converts the feature map into a feature sequence along its horizontal axis. The feature sequence is in the form of a sequence of vectors. After that, ASTER uses a multi-layer bidirectional LSTM[26] (BLSTM) network over the feature sequence.

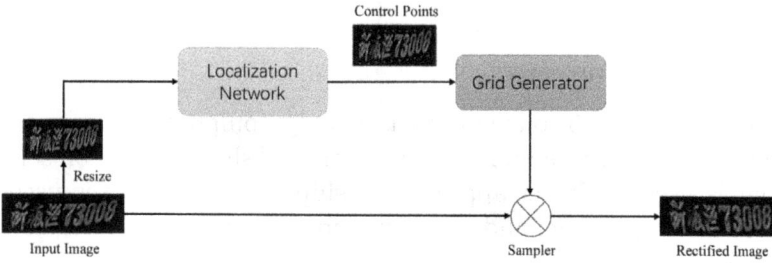

Fig. 3:   Rectification network.

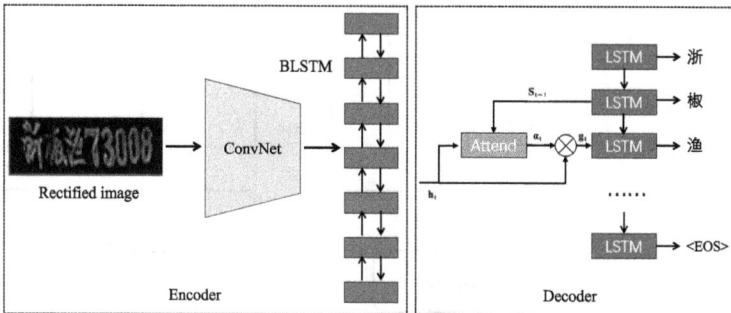

Fig. 4:   The text recognition network.

The sequence-to-sequence model can transfer a feature sequence into a character sequence. According to the name of a ship license plate, the length of text is variable in the sequence-to-sequence model. The decoder is based on the attention sequence-to-sequence model. It uses the attention mechanism at the prediction stage. Attention mechanism means paying attention to other words in a text while predicting each word. An attention-based sequence-to-sequence model is typically one directional, from left to right. ASTER proposes a bidirectional approach. It consists of two decoders with opposite directions. The better one is selected.

The following are our setups. The size of the input image of the localization network is 32×64 and the size of the rectification output is 32×100. The number of control points is set to 20. The CNN structure has six convolutional layers inserted with maxpool layers using a 2×2 kernel followed by two linear layers. Its output size is 40, which is twice of the control points number.

5.   Experiment

5.1.   *System overview*

As shown in Fig. 5, our system is mainly composed of four parts: radar, video camera, ship location server, and ship recognition server. Radar is used to detect approaching ships; video camera is used to collect the ship image; and ship detection server can get information

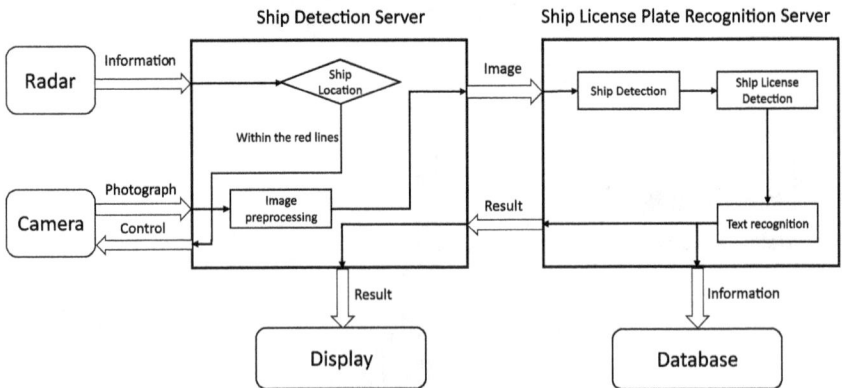

Fig. 5:   Components of ship license plate recognition system.

of the passing ship from a radar and control the video camera. It delivers images and ship information to the recognition server. Our algorithm is implemented in the recognition server. It operates the ship license plate detection, text rectification, and text recognition functions.

5.2. *Ship detection server*

Our system is deployed at the port and is used to detect and identify passing ships. In the initial stage, it deploys the radar to locate the ship and uses the camera to take pictures of the ship. When a ship approaches the port, the radar can detect the ship and send the location, speed, course, and other information to the server. The camera then takes photographs based on this information. The images are sent to the ship location server for further processing. The following are details of ship location work flow.

The radar and camera are shown in Fig. 6. The type of radar is MR1210. It is able to fit the change of temperature in port and its measurement range can cover the entire port. The type of camera is Hikvision iDS-2DYH4FFW-NZL. It can adapt to the wind and rain environment of the port, support night detection, and dynamically capture the movement of ships and follow them. The radar detection process is shown in Fig. 7. Water area around the port is divided into several parts. If no ship travels to the area surrounded by blue

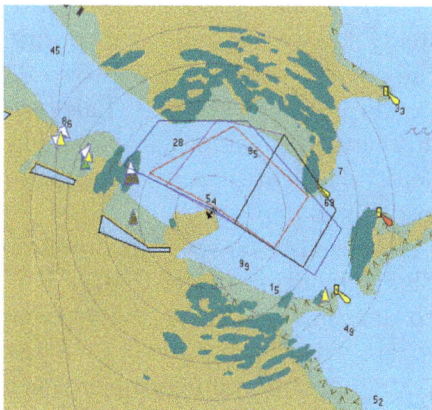

Fig. 6:   Radar monitoring map.

Fig. 7: The devices we use in ship location stage. The left is camera and the right is radar.

lines, the server read frequency to radar is set to 1 Hz. If the ship travels to the area surrounded by blue lines, the server read frequency increases to 4 Hz. When the ship travels to the area surrounded by red lines, the server controls the camera to take picture of the ship. According to the radar feedback distance, cameras adjust the focal length automatically and capture the image of the ship. Then, images are transmitted to the ship plate recognition server for subsequent processing.

When the image of the ship is taken, the detection server sends the image, image shape, and number of images to the ship recognition system.

5.3. *Ship recognition server*

In the real world deployment, the distance between the ship and the camera is usually far even though the position of the ship is detected. When the camera begins moving toward the ship, usually only the bow of the ship can be captured. This condition may result in the reduced efficiency of the ship plate area detection algorithm. Therefore, ship detection algorithm is added before the ship plate area detection algorithm is deployed.

Faster R-CNN is utilized to detect ships from an image. When an image is sent to the recognition server, Faster R-CNN algorithm is carried out first to determine whether images contain ships or not. Afterwards, if there is a ship in the image, it will be transferred to the subsequent ship plate detection and ship plate recognition network. In the ship plate detection stage, the specific location of the ship plate is identified in the image of the ship, then the area of the ship plate is cut out. The picture of the ship plate is transferred to the ship plate recognition network, and the ship plate recognition network recognizes the ship plate. Finally, the ship license is matched with the database and the final result is returned.

The deployment of database checking is mainly to prevent the wrong result after the identification of the outgoing ship's license. We record the ship in the database when the ship enters the port. When the ship is detected to leave the port, we match it in the database. If the ship information exists, the ship will be recorded as having left the port.

5.4. *Evaluation metrics*

We use the same algorithm evaluation metrics in existing target detection tasks such as PASCAL object detection.[27] Considering the characteristics of the ship license plate detection task, we utilize metrics including precision, recall, F1 score, and average precision (AP). They are calculated as follows:

$$P = \frac{TP}{TP + FP}, \tag{2}$$

$$R = \frac{TP}{TP + FN}, \tag{3}$$

$$F1 = \frac{2 \times P \times R}{P + R}. \tag{4}$$

To calculate AP, assuming that there are M positive examples in N samples, we get M recall values. For each recall value r, we can calculate the maximal precision corresponding to $(r' > r)$, and then average the M precision values to get the final AP value. Generally, a classifier with higher AP performs better.

Since bounding boxes detected by the algorithm cannot completely match the manually labeled data, we use IoU (Intersect over Union) to evaluate the positioning accuracy of the bounding box. It is defined to measure the overlap of two boxes. The calculation is as follows:

$$IoU\,(A, B) = \frac{A \cap B}{A \cup B}. \tag{5}$$

Unlike the target detection that some detected object features are sufficient for subsequent recognition, the ship plate detection detects all characters in an image. Therefore, we set that when $IoU > 0.8$, the classification is correct, and the ship plate is regarded as a successful sample.

We use accuracy and average edit distance (AED)[28] as our evaluation metrics for ship license plate recognition. For an input ship license plate image, S_g is defined as the sequence string of the ground truth label of the ship license image, i.e., the true name of the ship license plate. We assume S_p as the output of the ship plate recognition algorithm, i.e., the predicted ship plate name for any two S_g and S_p. The minimal number of editing operations required to convert one string to another is called edit distance. The larger the edit distance, the lower the similarity between the two strings. The editing operations include the insertion, deletion, and replacement of characters.

Accuracy is calculated by comparing S_g and S_p directly. If the character length of S_p is equal to S_g, and the character of each position of S_p is the same as S_g, then S_p is considered to be a correct prediction. Otherwise, it is a wrong prediction.

The corresponding formulas are as follows:

$$AED = \frac{\sum_{i=1}^{N} ED\left(S_{g_i}, S_{p_i}\right)}{N}, \tag{6}$$

$$ACC = \frac{N_{\text{correct}}}{N} \times 100\%, \tag{7}$$

where ED is the edit distance, N is the number of samples, and N_{correct} is the number of correct recognition.

5.5. Evaluation and performance

In the experiment, we train the ship license plate detection model and the ship license plate recognition model separately. They are trained using GeForceGTX 1080Ti.

In the ship license plate detection stage, the images are resized to 1280×720 from 1920×1080. We enlarge the training set to 7,312 through horizontal flipping and data enhancement. The training set and testing set are divided by a ratio of 9:1. We train our detection model using stochastic gradient descent with momentum. The training iterations is 100,000. The learning rate is set to 0.0003. It decreases to 0.00003 and 0.000003, respectively, when trained to 30,000 iterations and 60,000 iterations.

As mentioned in the previous section, at detection stage of our task, high IoU are necessary for text recognition. We set the classification correct if $IoU > 0.8$. We evaluate the performance of our detection network in different λ_1 and λ_2 in Eq. (1). When setting $\lambda_1 = 1$ and $\lambda_2 = 0$ at training, it means we only regress axis-aligned bounding boxes. We set $\lambda_1 = 0$ and $\lambda_2 = 1$ to test the regression accuracy without axis-aligned bounding boxes. As shown in Table 2, the network with two regression branches performs better than the others.

We also compare our method with Faster R-CNN method and CTPN[16] in the same experiment settings. The performance is shown in Fig. 8. Our network outperforms both Faster R-CNN and CTPN in the ship license plate detection task.

Table 2: Running speeds of different methods.

Algorithm	Time (s)
Faster R-CNN	0.001
CTPN	0.144
Our method	0.128

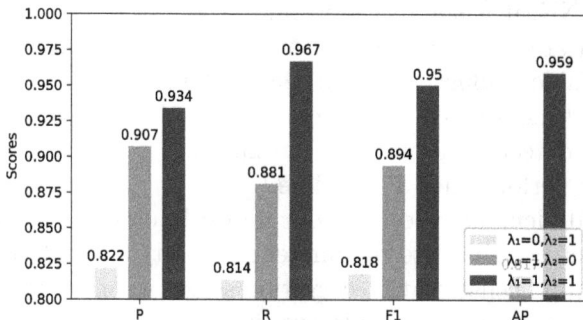

Fig. 8: Performance of ship license plate detection.

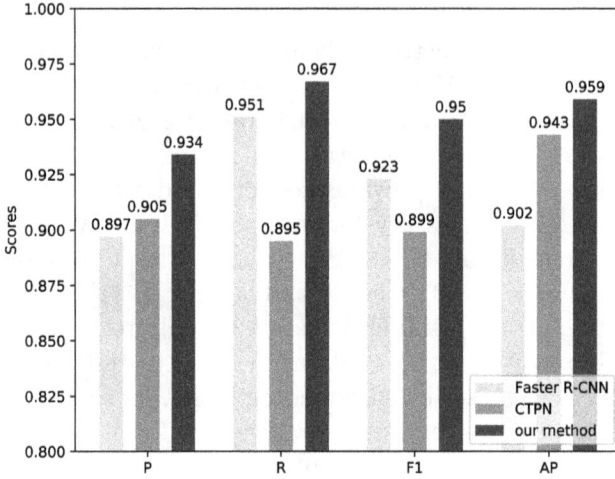

Fig. 9: Detection results of different methods.

Table 3: Performance of graying image.

Data	Daytime	Night
Normal data	0.960	0.869
Grayed data	0.929	0.934

We further test the mean processing time of every image, as shown in Fig. 9. Though the processing speed of our method is slower than Faster R-CNN, it is sufficient for our system.

We separate images at daytime and nighttime to test our detection network. Table 3 illustrates that the performance of detecting ship plate in daytime is much better than that in night images. By graying images, the detection accuracy at daytime and night are close, though it sacrifices performance at daytime.

Figure 10 demonstrates that our network for ship plate detection is able to detect ship plates in different visibility and light conditions.

In the ship license plate recognition stage, we use a total of 3,656 ship license plate images in different sizes. The are resized to 256×64 before entering rectification network and the training set and testing set are divided by the ratio of 9:1. The optimizer is ADADELTA.[29]

Fig. 10:   Some examples of detection results.

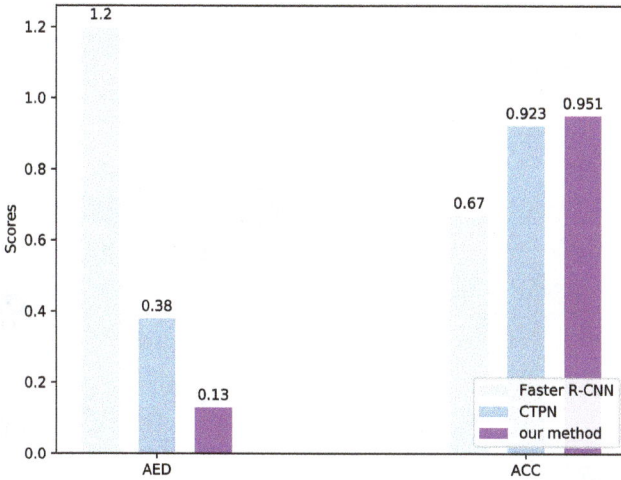

Fig. 11:   The recognition performance.

The training iterations is 50,000. The model is trained by batches of 32 examples for 200 epochs. The learning rate is set to 1.0 initially and decayed to 0.1 and 0.01 at 20,000 iterations and 40,000 iterations, respectively.

Figure 11 shows that our method can recognize the ship license plate with high accuracy. The Faster R-CNN's performance is not

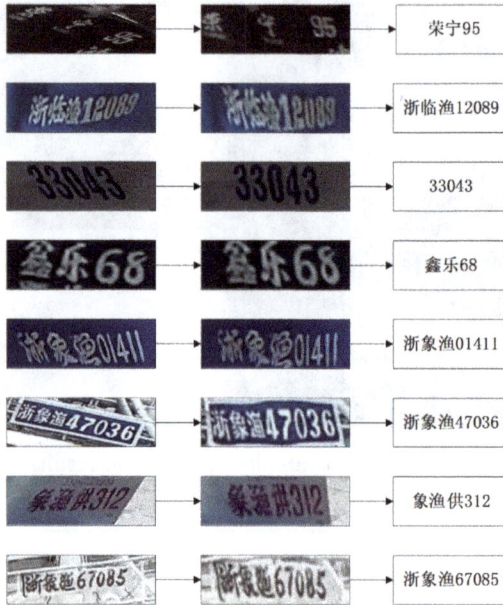

Fig. 12:   Some examples of recognition results.

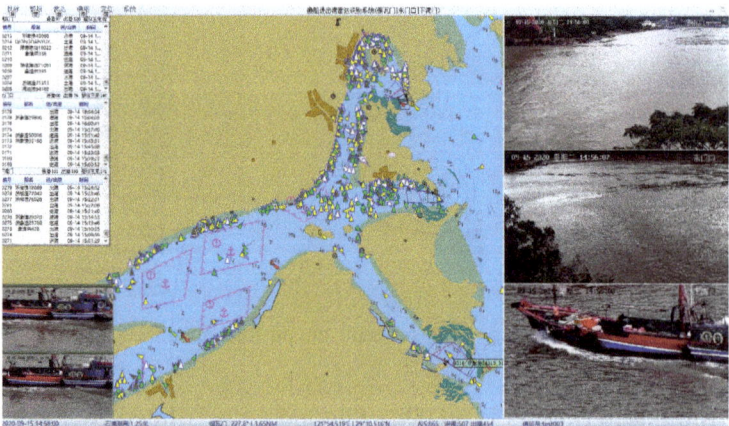

Fig. 13:   Running of our system.

good at recognition task, since it may consider one character as the whole text.

Figure 12 shows the rectification and recognition results. The left one is the input image; the middle one is the rectified image; and the right one is the recognition result. The rectification network we use can effectively rectify the irregular ship license plate, e.g., inclined ship license plate.

The client interface is shown in Fig. 13. The interface contains four parts: ship entry and exit port information, captured images, radar monitoring map, and port camera video stream information. In the radar detection image, each arrow represents the position of a ship, and the direction of the arrow is the direction of the ship. The port video stream information is on the right side of the image.

6. Conclusion

Deployment of cameras in ports makes it possible to use a vision-based method to identify ships. Considering the difficulty in recognizing ship license plates, a new method is proposed in this chapter. It is mainly composed of three stages including text detection, text rectification, and text recognition. We deploy this method in a ship recognition system at the port. It achieves high performance and can identify ships in real time.

References

1. A. Harati-Mokhtari, A. Wall, P. Brooks, and J. Wang, Automatic identification system (AIS): Data reliability and human error implications, *The Journal of Navigation.* **60**(3), 373 (2007).
2. Y. yang and C.-L. Zhu, License plate recognition and RFID techniques for design of community vehicle management system, *Journal of Jiangxi Vocational and Technical College of Electricity.* **97**(8), 31–33 (2012).
3. K. Yamaguchi, Y. Nagaya, K. Ueda, H. Nemoto, and M. Nakagawa, A method for identifying specific vehicles using template matching, in *Proceedings 199 IEEE/IEEJ/JSAI International Conference on Intelligent Transportation Systems* (Cat. No.99TH8383), 1999, pp. 8–13. doi: 10.1109/ITSC.1999.821019.
4. N. Dalal and B. Triggs, Histograms of oriented gradients for human detection, in *2005 IEEE Computer Society Conference on Computer Vision and Pattern Recognition (CVPR'05)*, 2005, vol. 1, pp. 886–893.
5. P. F. Felzenszwalb, R. B. Girshick, D. McAllester, and D. Ramanan, Object detection with discriminatively trained part-based models,

IEEE Transactions on Pattern Analysis and Machine Intelligence. **32**(9), 1627–1645 (2009).

6. P. Viola and M. Jones, Robust real-time face detection, *International Journal of Computer Vision.* **57**(2), 137–154 (2004).

7. W. Liu, D. Anguelov, D. Erhan, C. Szegedy, S. Reed, C.-Y. Fu, and A. C. Berg, SSD: Single shot multibox detector, in *European Conference on Computer Vision*, 2016, pp. 21–37.

8. C.-Y. Fu, W. Liu, A. Ranga, A. Tyagi, and A. C. Berg, DSSD: Deconvolutional single shot detector, arXiv preprint arXiv:1701.06659 (2017).

9. J. Redmon, S. Divvala, R. Girshick, and A. Farhadi, You only look once: Unified, real-time object detection, in *Proceedings of the IEEE Conference on Computer Vision and Pattern Recognition*, 2016, pp. 779–788.

10. J. Redmon and A. Farhadi, YOLO9000: Better, faster, stronger, in *Proceedings of the IEEE Conference on Computer Vision and Pattern Recognition*, 2017, pp. 7263–7271.

11. R. Girshick, J. Donahue, T. Darrell, and J. Malik, Rich feature hierarchies for accurate object detection and semantic segmentation, in *Proceedings of the IEEE Conference on Computer Vision and Pattern Recognition*, 2014, pp. 580–587.

12. P. Purkait, C. Zhao, and C. Zach, SPP-NET: Deep absolute pose regression with synthetic views, arXiv preprint arXiv:1712.03452 (2017).

13. R. Girshick. Fast R-CNN, in *Proceedings of the IEEE International Conference on Computer Vision*, 2015, pp. 1440–1448.

14. S. Ren, K. He, R. Girshick, and J. Sun, Faster R-CNN: Towards real-time object detection with region proposal networks, in *Advances in Neural Information Processing Systems*, 2015, pp. 91–99.

15. X. Zhou, C. Yao, H. Wen, Y. Wang, S. Zhou, W. He, and J. Liang, East: An efficient and accurate scene text detector, in *Proceedings of the IEEE Conference on Computer Vision and Pattern Recognition*, 2017, pp. 5551–5560.

16. Z. Tian, W. Huang, T. He, P. He, and Y. Qiao, Detecting text in natural image with connectionist text proposal network, in *European Conference on Computer Vision*, 2016, pp. 56–72.

17. Y. Liu and L. Jin, Deep matching prior network: Toward tighter multi-oriented text detection, in *Proceedings of the IEEE Conference on Computer Vision and Pattern Recognition*, 2017, pp. 1962–1969.

18. B. Shi, X. Bai, and S. Belongie, Detecting oriented text in natural images by linking segments, in *Proceedings of the IEEE Conference on Computer Vision and Pattern Recognition*, 2017, pp. 2550–2558.

19. M. Liao, B. Shi, X. Bai, X. Wang, and W. Liu, Textboxes: A fast text detector with a single deep neural network, arXiv preprint arXiv:1611.06779 (2016).

20. J. Ma, W. Shao, H. Ye, L. Wang, H. Wang, Y. Zheng, and X. Xue, Arbitrary-oriented scene text detection via rotation proposals, *IEEE Transactions on Multimedia*. **20**(11), 3111–3122 (2018).

21. Y. Jiang, X. Zhu, X. Wang, S. Yang, W. Li, H. Wang, P. Fu, and Z. Luo, R2CNN: Rotational region CNN for orientation robust scene text detection, arXiv preprint arXiv:1706.09579 (2017).

22. J. R. Uijlings, K. E. Van De Sande, T. Gevers, and A. W. Smeulders, Selective search for object recognition, *International Journal of Computer Vision*. **104**(2), 154–171 (2013).

23. K. He, X. Zhang, S. Ren, and J. Sun, Deep residual learning for image recognition, in *Proceedings of the IEEE Conference on Computer Vision and Pattern Recognition*, 2016, pp. 770–778.

24. B. Shi, M. Yang, X. Wang, P. Lyu, C. Yao, and X. Bai, Aster: An attentional scene text recognizer with flexible rectification, *IEEE Transactions on Pattern Analysis and Machine Intelligence*. **41**(9), 2035–2048 (2018).

25. F. L. Bookstein, Principal warps: Thin-plate splines and the decomposition of deformations, *IEEE Transactions on Pattern Analysis and Machine Intelligence*. **11**(6), 567–585 (1989).

26. A. Graves, M. Liwicki, S. Fernández, R. Bertolami, H. Bunke, and J. Schmidhuber, A novel connectionist system for unconstrained handwriting recognition, *IEEE Transactions on Pattern Analysis and Machine Intelligence*. **31**(5), 855–868 (2008).

27. M. Everingham, L. Van Gool, C. K. Williams, J. Winn, and A. Zisserman, The Pascal visual object classes (VOC) challenge, *International Journal of Computer Vision*. **88**(2), 303–338 (2010).

28. G. Navarro, A guided tour to approximate string matching, *ACM Computing Surveys (CSUR)*. **33**(1), 31–88 (2001).

29. M. D. Zeiler, Adadelta: An adaptive learning rate method, arXiv preprint arXiv:1212.5701 (2012).

https://doi.org/10.1142/9789811266911_0003

Chapter 3

Generative Adversial Network Enhanced Bearing Roller Defect Detection and Segmentation

Jiafei Shao, Zuohui Chen, and Qi Xuan[*]

Institute of Cyberspace Security,
Zhejiang University of Technology,
University of Nottingham Ningbo China,
Hangzhou, P.R. China

[*]*xuanqi@zjut.edu.cn*

As the main components of bearings, bearing rollers are for support and rotation. Thus, their quality directly determines the bearing performance. During the manufacturing of bearing rollers, various defects appear on the surface, which can severely affect the normal operation of the machine. In this chapter, we propose two methods with different detection granularity to identify the surface defects. The first one uses a binary classification network to detect defects and a generative adversarial network (GAN) for data augmentation. The other one focuses on detecting fine-grained defects, i.e., subtle scratches, which is based on the detection result of the first method. Experimental results show that our first technique improves the detection accuracy from 83.6% to 89.2% (average on side, end face, and chamfer images). The second one achieves accuracy of 93.8% on fine-grained defects.

1.   Introduction

Bearing is one of the essential components in most modern machinery.[1] It can ease the friction of mechanical structures and reduce mechanical loss. It is widely used in aviation, navigation, military industry, and many other fields. The quality of bearings directly affects the efficiency of machine operation. Several surveys on the causes of industrial machinery failures conducted by IEEE Industry Application Society (IEEE-IAS)[2–4] and the Japan Electrical Manufacturers' Association (JEMA)[5] show that bearing fault is the most common machine failure,[6] which accounts for 30% to 40%.

During the use of bearings, bearings with defects, such as scratches, scrapes, and markings, can severely interfere with the normal operation of the machine.[7] Currently, there are a variety of methods to detect surface defects of bearing rollers. They can be categorized into contact and non-contact.[8] Contact detection methods, such as infrared flaw detection, are highly sensitive to the external environment and require precise operation. Infrared flaw detector also suffers from scratching the bearing surface and poor detection efficiency. Traditional non-contact detection method, such as manual detection, is inefficient. For example, the ultrasonic detection method[9] is very expensive. The photoelectric detection method has a good adaptability to different bearing rollers, but is sensitive to light. The magnetic particle detection method[10] overcomes the above problems and can effectively detect the internal defects. However, this approach has security risks and cannot be applied in the production process.

Compared with the above non-contact detection methods, vision-based methods are efficient and inexpensive. Traditional vision-based methods require manually designed features, such as K-means algorithm,[11] support vector machine (SVM) linear classifier,[12–14] linear gray scale, and OSTU threshold segmentation,[15] etc. However, manually designed features are limited to a certain product and lack flexibility.[16] Compared with the manual features, deep learning can directly learn features from low-level data and exhibits great performance in many computer vision tasks. Therefore, we utilize a deep learning model to detect defects of bearing rollers.

A deep learning model requires massive data to achieve a good performance. However, there are no known public bearing roller defects datasets. The insufficient training data can damage the classification accuracy. To address the issue, we introduce GAN for data enhancement. We verify the validity of data augmentation through experiments. The main contents of this chapter are as follows:

- We evaluate four most popular convolutional neural networks (CNN) to detect the defects of bearing rollers. The best one achieves average accuracy of 83.6% on side, end face, and chamfer images (without data augmentation). To further improve the detection accuracy, we use a DCGAN to augment the original dataset. The detection accuracy increases to 89.2% with data augmentation.
- We use object detection networks to identify fine-grained defects based on the result of classification. The experimental results show that its average precision (0.938) and recall (0.940) surpass the first method.

The rest of the chapter is organized as follow. In Section 2, we introduce related works about object detection network, generative adversarial network, and related evaluation indicators. In Section 3, we introduce the overall process and the bearing roller defect detection system. In Section 4, we present experimental results and discussion. Finally, Section 5 holds our conclusions and future outlook.

2. Related Work

2.1. *Object detection network*

Object detection deals with detecting instances of visual objects, including object category and location.[17] Traditional object detection methods are based on handcrafted features and shallow trainable architectures,[18] such as CascadeClassifier,[19] histogram of oriented gradients (HOG),[20] and SVM,[12-14] etc. They combine multiple low-level image features with high-level context from object detectors and scene classifiers by constructing complex ensembles.[18]

However, manually designed features cannot adapt to new situations when the production lines and products are changed. Deep learning models have a better generalizability. Deep learning-based object detection algorithms can be divided into two categories: one-stage detection and two-stage detection, respectively.

The first one only uses one network, whose output is the target category probability and position coordinate. The mainstream one-stage detection algorithms include SSD[21] and YOLO,[22] etc. Although they are fast, their accuracy is usually lower than the two-stage ones. The two-stage detection algorithms first extract multiple candidate regions containing the target, and then perform region classification and location refinement.

Their representative algorithms include region proposal with convolutional neural network (R-CNN),[23] Fast R-CNN,[24] and Faster R-CNN,[25] etc. To ensure defect detection accuracy, we use two-stage detection method.

R-CNN[23] first extracts a set of independent region proposals (object candidate boxes) by selective search.[26] Then each region proposal is scaled to a fixed-size image and fed into a CNN model (e.g., AlexNet) to extract features. In the end, a linear SVM classifier is used to predict and classify the object in each region. For public datasets, R-CNN achieves a mean average precision (mAP) of 53.7% on PASCAL VOC 2010.[23] In 2015, Ross Girshick modified R-CNN and proposed Fast R-CNN. Fast R-CNN trains both detector and bounding box regressor simultaneously, which makes it capable of using very deep detection networks. It is up to nine times faster than R-CNN. However, both R-CNN and Fast R-CNN are based on the region proposal algorithms, which is very time-consuming. Ren et al. propose Faster R-CNN. It replaces the previous regional proposal method by using the regional proposal network (RPN). RPN is a fully convolutional network that takes images of any size as inputs and outputs a set of rectangular candidates. Anchor is the core concept of RPN network. An anchor point is the central point of the current slide window on the feature map generated by the backbone network, which will be further processed as a candidate box.

2.2. Generative adversarial network

Goodfellow et al.[27] are inspired by the min-max two-person zero-sum game and propose an adversarial process to train generative models. As an emerging unsupervised algorithm, generative adversarial networks train a pair of competing networks.[28,29] The two networks competing in the GAN are discriminator and generator. The discriminator is used to determine a sample belonging to a fake distribution or a real distribution. Its output is the probability that the sample is real. The generator deceives the discriminator by generating a fake sample. The two parties continue to compete in the iteration process, so that the generated images gradually approach the original images. As the study of GAN advances, many architectures are proposed. The first GAN architecture[27] used a fully connected network as the generator and discriminator and was applied to simple image datasets, such as MNIST. Mirza *et al.*[30] propose conditional GAN, where they add additional conditional information to the inputs. Radford *et al.*[31] propose deep convolutional generative adversarial networks (DCGANs), which have certain architectural constraints. Makhzani *et al.*[32] propose the adversarial autoencoder (AAE). It uses GAN to perform variational inference by matching the aggregated posterior of the autoencoder with an arbitrary prior distribution. Donahue et al.[33] propose bidirectional generative adversarial networks (BiGANs), which learn the inverse mapping from data to the latent space. Its resulting learned feature representation is useful for auxiliary supervised discrimination tasks.

2.3.   *The evaluation indicators*

In our first method, the bearing roller defect detection is a binary classification problem. Hence, we use the basic evaluation metrics of the binary classifier, including precision (P), recall (R), F1-score ($F1$), and area under the ROC curve (AUC).

Assuming TF is the positive samples correctly classified, FN is the positive samples wrongly classified, FP is the negative samples correctly classified, and TN is the negative samples wrongly classified. The indicators are calculated as follows.

$$P = \frac{TP}{TP + FP}, \tag{1}$$

$$R = \frac{TP}{TP + FN}, \tag{2}$$

$$F1 = \frac{2PR}{P + R} = \frac{2TR}{2TP + FN + FP}, \tag{3}$$

$$AUC = \frac{\sum_{i \in positiveClass} rank_i - \frac{M(1+M)}{2}}{M \times N}, \tag{4}$$

where M is the number of positive samples, N is the number of negative samples, i is the ith positive sample, and $rank_i$ is the number of the ith sample.

The precision is calculated as the ratio between TP and FP, which measures the model accuracy. The recall is calculated as the ratio between TP and FN, which measures the model's ability to recognize positive samples. The $F1$ sore is the harmonic value of P and R. AUC is the area enclosed by the ROC curve within the square.

3.   Dataset

3.1.   *Analysis of bearing rollers defects*

Common defects include flaking, pitting, wear, fretting, and cracks, which appear on the sides, end faces, and chamfers. The description of various defects is given in Table 1.

The defects in different situations are shown in Fig. 1.

3.2.   *Dataset collection and analysis*

The data acquisition system utilizes industrial light sources to construct the lighting environment and uses industrial cameras and lenses to acquire images. We choose a line scan camera LA-GM-04K08A to photograph the side, and an area scan camera acA2500-14gm to photograph the chamfer and the end face. As for the lens, we

Table 1:   Description of various bearing roller failures.

Failure	Description
Scratches	Defects caused by high contact pressure and heat on the rolling surface
Wear	Defects caused by sliding abrasion on parts, which is manifested as irregular shapes and large areas
Corrosion	Defects caused by oxidation or dissolution on the surface of bearing rollers
Lack of material	Due to the recesses on the chamfering, the roller profile is recessed
Undergrinding	Defects caused by insufficient grinding
Underpunching	Defects caused by insufficient punching

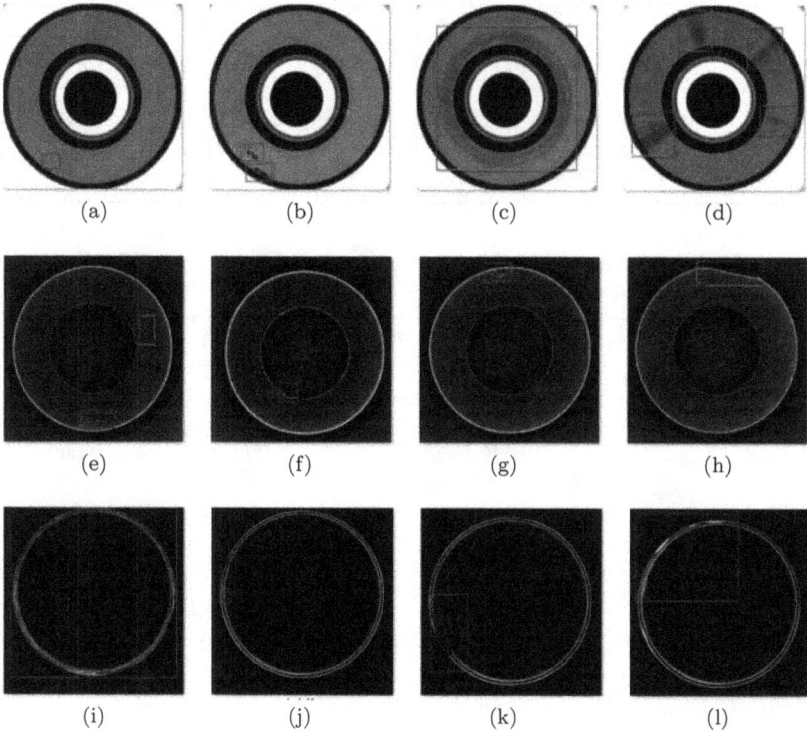

Fig. 1: Main types of defects on the side, end face, and chamfering. (a) to (d) show that the defects on the side are mainly scratches, wear, burns, and discoloration; (e) to (h) show that the defects on the end face are mainly scratches, inner diameter sags, wear, and outer diameter corner crack; (i) to (l) show that the defects on the chamfering are corrosion, scratches, broken corners, and wear.

use M1614-MP2 as the area scan camera lens and MK-APO-CPN as the line scan camera lens. LEDs have advantages in many aspects, such as reasonable price, good stability, and durability. Therefore, they are selected as the light sources of our data acquisition system.

A total of 2,912 original bearing roller images are collected from the bearing roller defect detection system. According to the camera angles, these images are divided into three categories, i.e., side, end face, and chamfer. There are 1,394 images in the side category, 788 images in the end face category, and 730 images in the chamfer category. We set the size of each image to 224×224. However, a few thousand images cannot guarantee high accuracy. We use DCGAN for the dataset augmentation. For each category, we randomly divide its

Table 2:   The original dataset of bearing roller defects.

Station	Original dataset	Training dataset	Test dataset	Proportion
Side	1,394	1,046	348	8:2
End face	788	588	200	8:2
Chamfer	730	548	182	8:2

Table 3:   Final training set of bearing roller defects.

Station	Original qualified samples	Original defective samples	Augmented qualified samples	Augmented defective samples
Side	523	523	2,092	2,092
End face	294	294	1,176	1,176
Chamfer	274	274	1,096	1,096

images into training set and testing set with a ratio of 8:2. The training set and testing set evenly contain qualified samples and defective samples. Finally, we quadruple each category in the training set by data augmentation. The original dataset is shown in Table 2 and the final augmented training set is shown in Table 3.

For our first method, i.e., the binary classification algorithm, we can use the augmented dataset directly. The second method requires data annotation, we use Labelme to annotate the scratched part.

4.   Method

The framework of our first method is shown in Fig. 2. It evaluates different networks and trains them with augmented data. The second method is shown in Fig. 3, which adds semantic segmentation modules on the basis of the previous method. It consists of three modules, where Module A is a faster R-CNN network, Module B is the binary classifier (the one evaluated in our first method), and Module C is a DeepLabV3+ network. Module A and B are the first stage of the detection process, which determines whether bearing rollers are

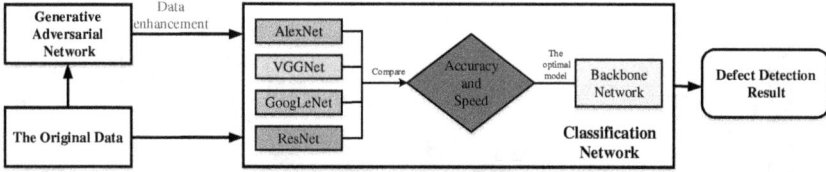

Fig. 2: The framework of the first defect detection method based on GAN. The classification network includes four basic neural network options, which are AlexNet, VGGNet, GoogLeNet, and ResNet.

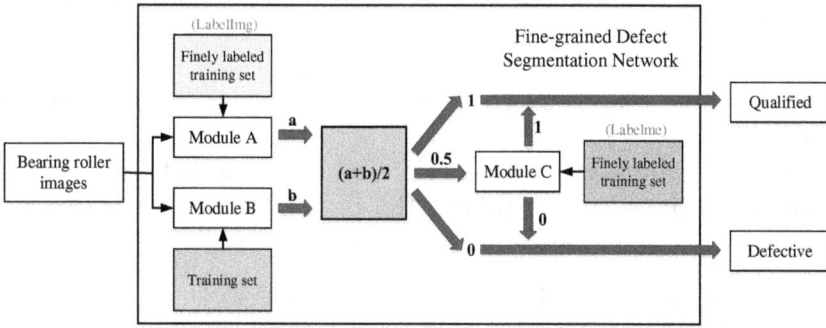

Fig. 3: The framework of the second defect detection method, which consists of two stages. In the first stage, Module A and Module B are independently trained, where Module A represents a Faster R-CNN network and Module B represents the binary classifier (the one evaluated in our first method. In the second stage, small scratches are detected through the DeepLabV3+ network.

defective. Module C further analyzes the results of the first stage to detect fine-grained defects.

4.1. *DCGAN network*

Compared with GAN, DCGAN has the following characteristics[29,31,34,35]:

- The pooling is replaced by the strided convolution to make the model downsample from the learning space.
- The generator utilizes the transposed convolution.
- Data preprocessing adopts batch normalization.
- Adopt a new gradient descent calculation method.

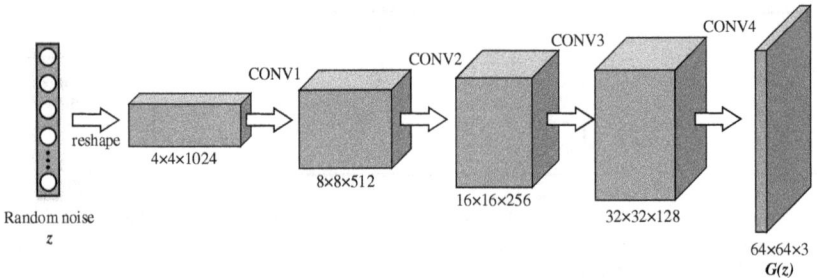

Fig. 4:   Architecture of DCGAN generator. A 100-dimensional uniform distribution z is projected into a series of fractionally-strided convolutions to form a 64×64 pixel image.

- Remove the fully connected layer in the overall structure of the model.
- The discriminator utilizes Leaky ReLU as the activation function, and the generator utilizes ReLU and Tanh as the activation functions.

The architecture of DCGAN is shown in Fig. 4. The input $z \in [-1, 1]^{1 \times 100}$ of the generator G is a random noise, which satisfies a uniform distribution. $G(z)$ is generated by G with 64×64 pixels. Then, we feed $G(z)$ and the real image x into the discriminator D, and D output probability of the real image.

In this work, the data augmentation contains both classical augmentation and DCGAN-based augmentation. For classical data augmentation, we crop these samples to the same size and rotate each image four times. For the DCGAN-based augmentation, we generate defect samples of each workstation.

4.2.   *Image classification network*

After the DCGAN model generates the defective samples, we use the enhanced dataset to train the classification networks and fine-tune parameters of the network to obtain the best detection performance. The optimal classification network is selected.

AlexNet[36]: The architecture of AlexNet network includes eight layers. The first five are convolutional layers and the last three are fully-connected layers. Each convolutional layer uses ReLU activation function. Response-normalization layers are used in the first, second, and third convolutional layers. Max-pooling layers follow the response-normalization layers and the fifth convolutional layer. Finally, the output of the last fully connected layers is sent to a softmax layer to generate the final result.

VGGNet[37]: VGGNet replaces large convolution kernels with smaller kernels and improves performance by deepening the network structure. Its structure includes convolution layers, maximum pooling layers, full connection layers, and a softmax output layer. All convolutional layers in VGGNet use 3 × 3 convolution kernels and the padding is set to 1. Each convolution layer is followed by a max-pooling layer. The last three layers are fully connected layers, where the first two layers have 4,096 channels and the last one has two channels. The activation functions of all hidden layers are ReLU, and the last layer is a softmax layer.

GoogLeNet[38]: GoogLeNet (Inception) uses filter sizes 1 × 1, 3 × 3, and 5 × 5 to reduce computational complexity. The visual information is processed at various scales and then aggregated. In this structure, although the network depth is 22 layers, the amount of parameters is 1/12 of AlexNet. *ResNet*[20]: ResNet uses residual connection to facilitate the training of networks. It avoids gradient explosion and gradient disappearance caused by network deepening. ResNet consists of residual blocks. In a residual block, there are two paths to the output, including a shortcut connection and convolutional layers. The shortcut connection, or residual connection, makes it easier for gradients to reach shallow neurons. Thus, we can train very deep networks.

4.3. *Fine-grained defect segmentation network*

As shown in Fig. 3, this network is composed of three modules which are trained using different data. Module A is a Faster R-CNN network; module B is the binary classification network used in our first method; and Module C is a DeepLabV3+ network.

Faster R-CNN network combines a region proposal network (RPN) and an object detection network fast R-CNN.[38,39] As shown in Fig. 5, an image is fed into the backbone network for CNN feature extraction, which contains 13 convolution layers and 4 pooling layers. Then we feed the feature map into RPN network, which provides foreground discrimination and candidate frame positioning.

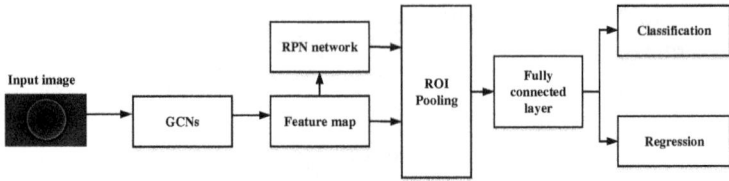

Fig. 5: The structure of Faster R-CNN network.

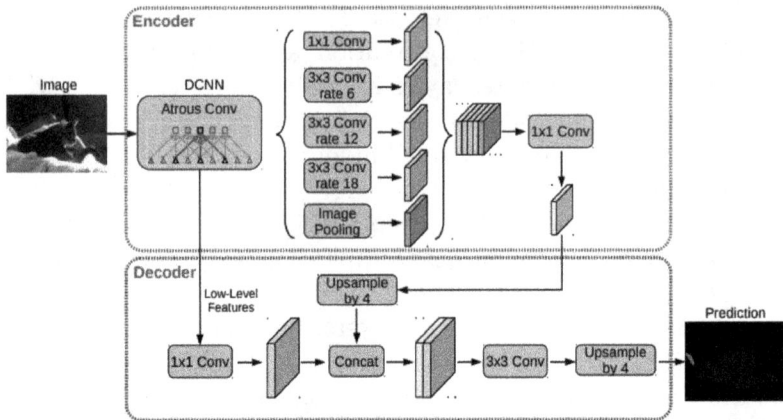

Fig. 6: DeepLabV3+ overall structure schematic. DeepLabV3+ network inherits DeepLabV3+ network, and employs a encoder-decoder structure to fuse multi-contextual information.[41]

To detect objects of different shapes and sizes, anchor mechanism is also used here. After obtaining region of interest (ROI), the object detection network gives the final classification result and defect locations.

DeepLabV3+ network employs an encoder-decoder structure to fuse multi-contextual information,[40,41] as shown in Fig. 6. The encoder module uses atrous spatial pyramid pooling (ASPP) with atrous convolution. Atrous convolution can capture contextual semantic information by adjusting the receptive field of the convolution filter. Therefore, it can extract multi-scale features. The decoder is responsible for restoring the scale of feature maps. To compensate the spatial detail information, the encoder merges its low-level features to improve segmentation accuracy. The high-level features are bilinearly upsampled four times and low-level features have dimensions reduced to fit high-level features. Finally, they are concatenated to obtain the final result. If there is defective area segmented in the final result, we classify the image as defective. Otherwise, it is qualified.

For our second method, input image is fed into Module A and B, to obtain the classification results, i.e., the probability of defectiveness. If both networks believe it is defective or qualified, the result is final. Otherwise, the image is fed into the semantic segmentation network for fine-grained classification.

5. Experimental Results

5.1. *Experimental setup*

The second method is composed of multiple modules, which require different data. As shown in Fig. 3, Module B is trained on the binary labeled data. We use 500 images of all the views from the training set and mark all defects to train Module A. Module C needs to detect small scratches. Thus, we mark the scratches in 200 images from the training set. We use *Tensorflow* as a deep learning framework in Python (version 3.6) and *Labelme* to mark the defects. The experimental environment setup is shown in Table 4.

We use Adam optimizer and cross-entropy loss to train our models. The learning rate and batch size are set to 0.001 and 64. The four evaluation indicators introduced in Section 2.3, i.e., precision (P), recall (R), $F1$-score $(F1)$, and area under the ROC curve (AUC), are used to evaluate our methods.

5.2. *Binary classification networks*

In this experiment, we use the original dataset to train the classification networks separately and select the best model. They are

Table 4: Workstation specification.

Hardware (machine type)	Software
CPU: Intel Core i7-6700	Ubuntu16.04 LTS
RAM: 16G	Python: 3.6
GPU: TITAN X(ascal)	Tensorflow-gpu-1.12.0

Table 5:　Performance of four classification networks (side).

Network	Precision	Recall	$F1$	AUC
AlexNet	0.793	0.815	0.794	0.797
VGGNet	0.805	0.778	0.790	0.801
GoogLeNet	0.801	0.837	0.810	0.805
ResNet	0.842	0.855	0.845	0.843

Table 6:　Performance of four classification networks (end face).

Network	Precision	Recall	$F1$	AUC
AlexNet	0.803	0.819	0.811	0.809
VGGNet	0.814	0.847	0.816	0.822
GoogLeNet	0.821	0.838	0.825	0.837
ResNet	0.863	0.859	0.860	0.861

Table 7:　Performance of four classification networks (chamfer).

Network	Precision	Recall	$F1$	AUC
AlexNet	0.783	0.791	0.785	0.779
VGGNet	0.791	0.782	0.784	0.781
GoogLeNet	0.798	0.787	0.791	0.797
ResNet	0.804	0.839	0.812	0.810

evaluated by accuracy (P, R, $F1$, and AUC) and speed (images per second). There are three views in our data, the result of different views are demonstrated separately. Results of training with original data are shown in Tables 5–7. Results demonstrate that ResNet has the best performance. ResNet achieves the best result in the end face, where the precision, recall, $F1$, and AUG are 0.863, 0.859, 0.860, and 0.861, respectively.

We also consider the detection speed when selecting the best model. As shown in Fig. 7, the detection speed of ResNet is also faster than other three networks. Hence, we select ResNet as the backbone network of the first defect detection method.

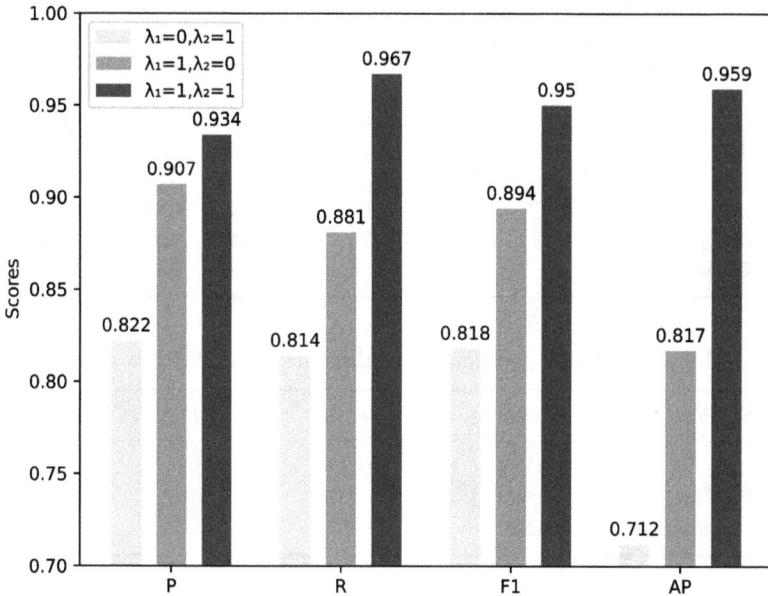

Fig. 7: Comparison of processing speed of four classification networks.

5.3. Data augmentation

In addition to the original training set, we add GAN generated data in different proportions to train the network and compare the detection performance. The proportions of new data are 0%, 10%, 20%, 30%, 40%, and 50%. The parameters of DCGAN are set to default, where the number of iterations is 500 and the learning rate is 0.0002.

As shown in Table 8, a proper amount of new data (less than 30%) improves detection accuracy. When the proportion of new data is 30%, the network detection performance is the best. When it is over 40%, new data can damage the performance. As shown in Table 9, the precision of the end face increases from 0.863 to 0.902, the recall increases from 0.859 to 0.924, F1 increases from 0.860 to 0.910, and AUC increases from 0.861 to 0.907.

5.4. Fine-grained detection

For our second method, we select Xception-65 as the backbone network in Module C. In the encoder part, we use atrous convolution

Table 8:   Influence of new data on network performance (end face).

Proportion of new data (%)	Precision	Recall	$F1$	AUC
0	0.863	0.859	0.860	0.861
10	0.871	0.882	0.875	0.873
20	0.885	0.897	0.890	0.896
30	0.902	0.924	0.910	0.917
40	0.875	0.894	0.879	0.883
50	0.858	0.834	0.846	0.852

Table 9:   Network performance at three stations (30% new data).

Station	Precision	Recall	F1	AUC
Side	0.889	0.913	0.892	0.903
End face	0.902	0.924	0.910	0.917
Chamfer	0.885	0.854	0.867	0.882

Table 10:   Detection performance of fine-grained defect segmentation network.

Station	Precision	Recall	F1	AUC
Side	0.936	0.951	0.941	0.947
End face	0.957	0.976	0.958	0.963
Chamfer	0.922	0.893	0.902	0.911

kernels with convolution rates of 6, 12, and 18 and implement the cross-entropy loss function.

Table 10 shows the detection performance of fine-grained defect segmentation network at different views. The segmentation network has the best performance in the end face, where the precision, recall, $F1$, and AUG are 0.957, 0.976, 0.958, and 0.963, respectively. The performance of the defect detection network is greatly improved.

The performance of the original model, data augmentation, and fine-grained segmentation are summarized in Table 11 (all use end face view). Overall, data augmentation based on DCGAN can effectively improve the performance of the original network. Fine-grained defect segmentation network has the highest accuracy, but its detection speed is slower than the binary classification networks.

Table 11: Performance comparison of two detection networks (end face).

Plan	Precision	Recall	$F1$	AUC	Speed
Based on classification network (no new data added)	0.863	0.859	0.860	0.861	—
Based on DCGAN (add 30% new data)	0.902	0.924	0.910	0.917	5
Based on fine-grained defect segmentation network	0.957	0.976	0.958	0.963	0.9

6. Conclusion

In the chapter, we show the application of data augmentation, object detection, and semantic segmentation in bearing roller defect detection. We propose two detection methods of different detection speed and granularity. The first one uses DCGAN to augment the training data, while the second method uses semantic segmentation module to further detect fine-grained defects which cannot be identified by the first method.

References

1. V. Sugumaran and K. I. Ramachandran, Effect of number of features on classification of roller bearing faults using SVM and PSVM, *Expert Systems with Applications.* **38**(4), 4088–4096 (2011).
2. Motor Reliability Working Group and others, Report of large motor reliability survey of industrial and commercial installations, Part I, *IEEE Transactions on Industrial Applications.* **1**, 865–872 (1985).
3. Motor Reliability Working Group and others, Report of large motor reliability survey of industrial and commercial installations, Part II, *IEEE Transactions on Industry Applications.* **21**(4), 853–864 (1985).
4. Motor Reliability Working Group and others, Report of large motor reliability survey of industrial and commercial installations, *IEEE Transactions on Industry Applications.* **23**(4), 153–158 (1987).
5. S. Zhang, B. Wang, M. Kanemaru, C. Lin, D. Liu, M. Miyoshi, K. Teo, and T. Habetler, Model-based analysis and quantification of bearing faults in induction machines, *IEEE Transactions on Industry Applications.* **56**, 2158–2170, (2020). doi: 10.1109/TIA.2020.2979383.

6. S. Zhang, S. Zhang, B. Wang, and T. G. Habetler, Deep learning algorithms for bearing fault diagnostics — A comprehensive review, *IEEE Access.* **8**, 29857–29881, (2020).

7. K. Adamsab, Machine learning algorithms for rotating machinery bearing fault diagnostics, *Materials Today: Proceedings.* **44**, 4931–4933 (01, 2021). doi: 10.1016/j.matpr.2020.12.050.

8. H. H. Bakker, R. C. Flemmer, and C. L. Flemmer, Coverage of apple surface for adequate machine vision inspection, in *2017 24th International Conference on Mechatronics and Machine Vision in Practice (M2VIP)*, 2017, pp. 1–5.

9. Z. Hui, G. Song, Y. Kang, and Z. Guan, A multi-channel on-line ultrasonic flaw detection system, *IEEE.* **2**, 5290–5293 (2006).

10. J. Luo, Z. Tian, and J. Yang. Fluorescent magnetic particle inspection device based on digital image processing, in *Intelligent Control & Automation*, 2015, pp. 5677–5681.

11. S. H. Jeong, H. T. Choi, and S. R. Kim, Detecting fabric defects with computer vision and fuzzy rule generation. Part I: Defect classification by image processing, *Textile Research Journal.* **71**(6), 518–526 (2001).

12. D. Mery and C. Arteta. Automatic defect recognition in x-ray testing using computer vision, in *2017 IEEE Winter Conference on Applications of Computer Vision (WACV)*, 2017, pp. 1026–1035.

13. H. Zhen and J. M. Parker. Texture defect detection using support vector machines with adaptive gabor wavelet features, in *Application of Computer Vision, 2005. WACV/MOTIONS '05 Volume 1. Seventh IEEE Workshops on*, 2005, pp. 275–280.

14. Z. Shen and J. Sun, Welding seam defect detection for canisters based on computer vision, in *2013 6th International Congress on Image and Signal Processing (CISP)*, 2013, pp. 788–793.

15. X. Yan, Application of computer vision in defect bar code detection, *Journal of Huaqiao University (Natural Science).* **38**, 109–112 (2017).

16. D. Tabernik, S. Šela, J. Skvarč, and D. Skočaj, Segmentation-based deep-learning approach for surface-defect detection, *Journal of Intelligent Manufacturing.* **31**(3), 759–776 (2020).

17. Z. Zou, Z. Shi, Y. Guo, and J. Ye, Object detection in 20 years: A survey, arXiv preprint arXiv:1905.05055 (2019).

18. Z.-Q. Zhao, P. Zheng, S.-t. Xu, and X. Wu, Object detection with deep learning: A review, *IEEE Transactions on Neural Networks and Learning Systems.* **30**(11), 3212–3232 (2019).

19. P. F. Felzenszwalb, R. B. Girshick, and D. McAllester, Cascade object detection with deformable part models, in *2010 IEEE Computer Society Conference on Computer Vision and Pattern Recognition*, 2010, pp. 2241–2248.

20. N. Dalal and B. Triggs. Histograms of oriented gradients for human detection, in *2005 IEEE Computer Society Conference on Computer Vision and Pattern Recognition (CVPR'05)*, 2005, vol. 1, pp. 886–893.

21. W. Liu, D. Anguelov, D. Erhan, C. Szegedy, S. Reed, C.-Y. Fu, and A. C. Berg. SSD: Single shot multibox detector, in *European Conference on Computer Vision*, 2016, pp. 21–37.

22. J. Redmon, S. Divvala, R. Girshick, and A. Farhadi, You only look once: Unified, real-time object detection, in *Proceedings of the IEEE Conference on Computer Vision and Pattern Recognition*, 2016, pp. 779–788.

23. R. Girshick, J. Donahue, T. Darrell, and J. Malik, Rich feature hierarchies for accurate object detection and semantic segmentation, in *Proceedings of the IEEE Conference on Computer Vision and Pattern Recognition*, 2014, pp. 580–587.

24. R. Girshick, Fast R-CNN, in *Proceedings of the IEEE International Conference on Computer Vision*, 2015, pp. 1440–1448.

25. S. Ren, K. He, R. Girshick, and J. Sun, Faster R-CNN: Towards real-time object detection with region proposal networks, *Advances in Neural Information Processing Systems*. **28**, 91–99 (2015).

26. K. E. Van de Sande, J. R. Uijlings, T. Gevers, and A. W. Smeulders, Segmentation as selective search for object recognition, in *2011 International Conference on Computer Vision*, 2011, pp. 1879–1886.

27. I. J. Goodfellow, J. Pouget-Abadie, M. Mirza, B. Xu, D. Warde-Farley, S. Ozair, A. Courville, and Y. Bengio, Generative adversarial networks, *Communications of the ACM*. **63**(11), 139–144 (2014).

28. H. Alqahtani, M. Kavakli-Thorne, and G. Kumar, Applications of generative adversarial networks (GANS): An updated review, *Archives of Computational Methods in Engineering*. **28**(2), 525–552 (2021).

29. A. Creswell, T. White, V. Dumoulin, K. Arulkumaran, B. Sengupta, and A. A. Bharath, Generative adversarial networks: An overview, *IEEE Signal Processing Magazine*. **35**(1), 53–65 (2018).

30. M. Mirza and S. Osindero, Conditional generative adversarial nets, arXiv preprint arXiv:1411.1784 (2014).

31. A. Radford, L. Metz, and S. Chintala, Unsupervised representation learning with deep convolutional generative adversarial networks, arXiv preprint arXiv:1511.06434 (2015).

32. A. Makhzani, J. Shlens, N. Jaitly, I. Goodfellow, and B. Frey, Adversarial autoencoders, arXiv preprint arXiv:1511.05644 (2015).

33. J. Donahue, P. Krähenbühl, and T. Darrell, Adversarial feature learning, arXiv preprint arXiv:1605.09782 (2016).

34. W. Fang, F. Zhang, V. S. Sheng, and Y. Ding, A method for improving CNN-based image recognition using DCGAN, *Computers, Materials and Continua*. **57**(1), 167–178 (2018).

35. Q. Wu, Y. Chen, and J. Meng, DCGAN-based data augmentation for tomato leaf disease identification, *IEEE Access.* **8**, 98716–98728 (2020).
36. A. Krizhevsky, I. Sutskever, and G. E. Hinton, Imagenet classification with deep convolutional neural networks, *Advances in Neural Information Processing Systems.* **25**, 1097–1105 (2012).
37. K. Simonyan and A. Zisserman, Very deep convolutional networks for large-scale image recognition, arXiv preprint arXiv:1409.1556 (2014).
38. C. Szegedy, W. Liu, Y. Jia, P. Sermanet, S. Reed, D. Anguelov, D. Erhan, V. Vanhoucke, and A. Rabinovich, Going deeper with convolutions, in *Proceedings of the IEEE Conference on Computer Vision and Pattern Recognition*, 2015, pp. 1–9.
39. S. Ren, K. He, R. Girshick, and J. Sun, Faster R-CNN: Towards real-time object detection with region proposal networks, *Advances in Neural Information Processing Systems.* **28**, 91–99 (2015).
40. A. R. Choudhury, R. Vanguri, S. R. Jambawalikar, and P. Kumar, Segmentation of brain tumors using DeepLabV3+, in *International MICCAI Brainlesion Workshop*, 2018, pp. 154–167.
41. L.-C. Chen, Y. Zhu, G. Papandreou, F. Schroff, and H. Adam. Encoder-decoder with atrous separable convolution for semantic image segmentation, in *Proceedings of the European Conference on Computer Vision (ECCV)*, 2018, pp. 801–818.

Chapter 4

Application of Deep Learning in Crop Stress

Qijun Chen, Qi Xuan, and Yun Xiang*

Institute of Cyberspace Security,
Zhejiang University of Technology,
Hangzhou, P.R. China

*xiangyun@zjut.edu.cn

For farmers, the variety of crop diseases and symptoms causes great difficulties in the diagnosis of plant diseases. To address this problem, researchers have recently begun using deep learning algorithms to analyze crop stress and achieved good results. This chapter aims to explore the research of deep learning in crop stress. To better illustrate the issue, this chapter begins with the types and challenges of crop stress, followed by an introduction to the deep neural networks used, and concludes with a summary of the current progresses, limitations, and future work of crop stress research.

1. Introduction

The world is paying increasing attention to smart agriculture,[1] which refers to leveraging advanced technology in the agricultural industry, realizing automated management, and increasing the quality and quantity of crops. However, an important challenge is how to process the collected images and structured monitoring data.[2]

Traditional processing techniques include machine learning methods (e.g., K-means clustering, support vector machines, artificial neural networks), linear polarization, and wavelet filtering, etc. However, they are all sensitive to data filtering and require massive labor on data processing. In recent years, with the rapid development of Deep Learning (DL) algorithms, intelligent agriculture based on big data and DL have become a new research focus, especially in the fields of pest detection, plant identification, crop and weed detection, and classification,[3,4] etc.

DL has better feature extraction capability compared with traditional processing technology. It achieves the state-of-the-art performance in image recognition,[5] object classification and detection,[6] face recognition[7] and speech recognition,[8] etc. The rapid development of embedded devices also provides rich data for developing smart agriculture.

In this section, we investigate and summarize representative crop stress papers from 2018 to 2021. We also provide a summary and outlook on the application of deep learning to crop stress.

2. A Mainstream Network of Deep Learning in the Field of Plants

Generally, farmers need classification, pheromone identification, and regional monitoring services for crop stress. The most popular deep learning networks include Convolutional Neural Networks (CNN), autoencoders, RCNN, and YOLO. This section introduces the above mainstream networks.

2.1. *Convolutional neural network*

Image classification is an important problem in computer vision. Traditional image classification methods depend on the hand-crafted features. There are many disadvantages such as difficulty in feature design, limitations for complex tasks, and difficulty in designing hierarchical relationships between features, which lead to weak generalization performance. With the emergence of large datasets and the development of data processing hardware, e.g., graphics processing unit (GPU), CNN is able to solve the above problems.[9]

A CNN consists of multiple layers, including an input layer, a series of hidden layers, and an output layer. The hidden layers of a standard CNN are mainly composed of convolutional layers, pooling layers, and fully connected layers. The objective of the convolutional layers is to extract features using convolution kernels. The pooling layer reduces the model dimensions by compressing the feature maps, which also improves the computational efficiency. The main function of the fully connected layer is to integrate the features and output the classification result.

In 2012, Krizhevsky *et al.*[10] proposed AlexNet, which achieved the best classification results in the ImageNet[11] Large-scale Visual Recognition Challenge Competition. Since then, CNN has become the focus of deep learning researchers and the mainstream architecture for most image recognition, classification, and detection tasks.[12] The most popular CNN-based networks include AlexNet,[10] Clarifai,[13] SPPNet,[14] VGGNet,[15] GoogleNet,[16] ResNet,[17] etc. In addition to the architecture, the study of CNN also promoted development in many related areas, such as parameter optimization strategies, neural optimization architecture (NOA), meta-learning, etc.

2.2. Autoencoder

The autoencoder was proposed by Hinton *et al.* in 2006,[18] which is an unsupervised learning method that uses the input data as supervision to learn a mapping relationship and obtain a reconstructed output. Its variants include sparse autoencoder, denoising autoencoder, and contract autoencoder, etc. They can be used for feature dimensionality reduction and feature extraction. A simple autoencoder can be structured as a three-layer neural network.

2.3. Object detection network

The most popular object detection networks include R-CNN,[19] OverFeat,[20] Fast/Faster R-CNN,[21,22] SSD,[23] and YOLO.[24-27] They can be divided into one-stage techniques and two-stage techniques.

2.3.1. *Two-stage methods*

Two-stage methods, such as the R-CNN models, start with candidate areas. Then they are classified. Ross Girshick *et al.*[19] proposed R-CNN in 2014, which has been ground-breaking in deep learning-based object detection. They applied convolutional neural networks to feature extraction, and the detection rate of PASCAL VOC dataset is 53.7%. In 2015, Girshick *et al.*[21] proposed Fast R-CNN, which performs feature extraction on the whole image and uses the ROI layer to process the features of the candidate regions. The ROI layer extracts features for each candidate region instead of the whole image. It greatly reduces the computational cost. Fast R-CNN uses multi-task loss to train the classification and regression networks simultaneously. In the same year, Ren *et al.*[22] proposed the Faster R-CNN network. The main contribution of Faster R-CNN is the Region Proposal Network (RPN), which uses the Anchor mechanism to combine the region generation and convolutional network. Its detection speed reaches 14 frames per second and the accuracy on VOC 2012 test set achieved 70.4%.

2.3.2. *One-stage Methods*

Though two-stage method has high accuracy, it is slow. One-stage methods, such as YOLO, can output detection results directly without finding candidate areas. The core idea of YOLO is to split an image into grids, and then determine the location of the target in each mesh to obtain the target's bounding box and the probability of its category. Redmon *et al.*[24] proposed YOLOv1 in 2015, and it is the pioneer of one-stage detection. They[25] also proposed the YOLOv2 network in 2017, which uses batch normalization, high resolution classifier, and anchor to obtain faster and more accurate models. In 2018, Redmon *et al.*[26] proposed YOLOv3, which uses the residual model Darknet-53 and FPN architecture to achieve multi-scale detection. YOLOv4[27] proposed many optimization strategies based on the original YOLO architecture. It adds a backbone structure to the Darknet-53, which contains five cross-stage partial (CSP) modules. They can enhance the learning capability of the network and reduce the training cost.

3. Deep Learning-Based Crop Stress Researches

3.1. *Existing literature*

The data sources and size of our surveyed papers are listed in Table 1. The most common data are images and self-collecting datasets, whereas image is the majority. The acquisition of the image dataset can be divided into self-collecting and public ones, respectively. Self-collecting image data are derived using remote sensing, ground cameras, camera-mounted drone aerial photography, hyperspectral imaging, near-infrared spectrometer, etc. Public datasets generally come from existing well-known public standard libraries, such as Oxford-17 flower, Oxford-102 flower, PlantVillage, ICL, Flavia, etc. Structured numerical data are mainly obtained by non-visual sensors. Constructing a high-quality crop stress dataset requires considerable effort since many stresses show no obvious symptoms at early stage. A number of studies show that identification of abiotic stress can use transfer learning to reduce training cost.

The number of papers surveyed for each year is shown in Fig. 1. It indicates that the research combined with deep learning and crop stress has increased rapidly in the past two years. There are 43 studies in 2020 and 2021, accounting for more than 72%.

3.2. *Biotic crop stress*

The growth of plants is affected by many environmental factors by which the plant faces a variety of biotic and abiotic stress. Generally, agricultural production is affected by different degrees

Table 1:   The source of the dataset.

Type	Source	Sample size
Image data	Public datasets	3,000~96,867
	Collect online	400~3,000
	Drone and aerial photography	10~5,200
	Camera	600~40,000
	Hyperspectral imaging	40~284

Number of Papers

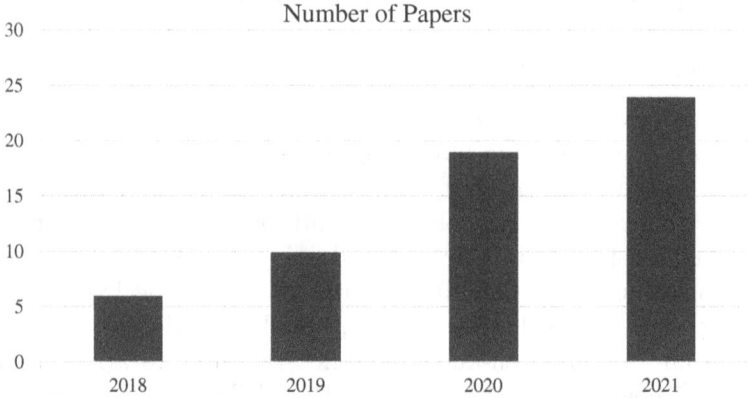

Fig. 1: The distribution of the year in which the selected papers were published.

of natural disasters. Once severe natural disasters occur, such as extreme high/low temperature, drought, floods, pests, and diseases, agricultural production suffers irreversible losses.[28] Drought, pests, and diseases are the main factors affecting world food production. The goal of intelligent agricultural plant protection is to identify the type and degree of stress, and to help farmers find the corresponding countermeasures.

Table 2 lists representative literature related to the recognition and identification of biotic stress in crops. The existing algorithms are mainly based on CNN, which can identify the type and degree of crop diseases through images.

Shin *et al.*[29] identified images of powdery mildew diseased leaves in strawberries. Strawberries in high temperature and humidity are prone to powdery mildew disease. In the early stages of infection, white mycelium begins to cover the leaf surface, which reduces photo-synthesis of strawberries and causes water shortage. As the infection develops, the white fungus group begins to cover the entire surface of the strawberry leaf, and the leaves become red-purple or have small purple spots. To protect the harvest, farmers need to identify the symptoms of powdery mildew disease efficiently and cure them in the early stages of infection. They showed that the CNN (ResNet-50) had a high accuracy (98.11%) in identifying powdery mildew disease. Moreover, considering the hardware limitation, the author proposed a lightweight network SqueezeNet-MOD2 with 92.61% accuracy.

Table 2: Application of deep learning in biotic stress of crops.

Application scenario	Author	Year	Algorithm
Powdery mildew disease detection on strawberry leaves	Shin et al.[29]	2021	CNN
Maize leaf disease classification	Ahila et al.[30]	2019	CNN
Tomato plant leaf disease detection	De et al.[31]	2018	F-RCNN
Betelvine leaf disease detection	Nickolas et al.[32]	2020	Mask-RCNN
Evaluating late blight severity in potato crops	Duarte et al.[33]	2018	CNN
Identification of disease and evaluation of bacteriosis in peach leaf	Yadav et al.[34]	2021	CNN
Recognition of apple leaf diseases	Tahir et al.[35]	2021	InceptionNet-V3
Wheat leaf rust detection	Azadbakht et al.[36]	2019	ν-SVR

Ahila et al.[30] used LeNet to identify corn diseases in the PlantVillage dataset (Fig. 2). They reduced the feature correlation by PCA albino preprocessing, and achieved an accuracy of 97.89%.

De et al.[31] constructed a dataset that contains 4,923 images of tomato plants. They classified tomato plant diseases (phoma rot, leaf miner, and target spot) based on this dataset, and established a system to detect tomato diseases (Fig. 3). The system achieved 95.75% accuracy in migrating learning disease identification models and 91.67% accuracy in practical applications. Nickolas et al.[32] used data collected online and labeled them manually to construct a dataset. They used a Mask-R-CNN to identify the disease of Betelvine leaves and achieved accuracy of 84%. Duarte et al.[33] used drones with cameras to capture multispectral data at the canopy of Betelvine. The multispectral data were then extracted by Agisoft Photoscanand. They used random forest to obtain an R^2 value of 0.75.

Yadav et al.[34] used CNN models to detect bacteria from peach leaf images. The images can be classified within 0.185s with an accuracy of 98.75%. Figure 4 shows the detection result on the peach leaf.

Fig. 2:   Sample images from PlantVillage dataset: (a) common rust, (b) gray leaf spot, (c) northern leaf blight, (d) healthy.

Note: Cited from Ahila *et al.* Maize leaf disease classification using deep convolutional neural networks.

Tahir *et al.*[35] trained a network based on Inception-V3 to identify apple leaf disease. They used transfer learning and retraining and achieved an accuracy of 97%. Azadbakht *et al.*[36] and Julio *et al.*[33] both made model predictions by collecting spectral data from the canopy. Azadbakht *et al.* added a leaf area index (LAI) to divide the crown data into three categories, i.e., high, medium, and low. They compared four machine learning methods to detect wheat leaf rust, including ν-support vector regression (ν-SVR), boosted regression trees (BRT), random forests regression (RFR), and Gaussian process regression (GPR). They found that the ν-SVR was superior to the other ML methods at all three leaf area levels, with R^2 index at around 0.99.

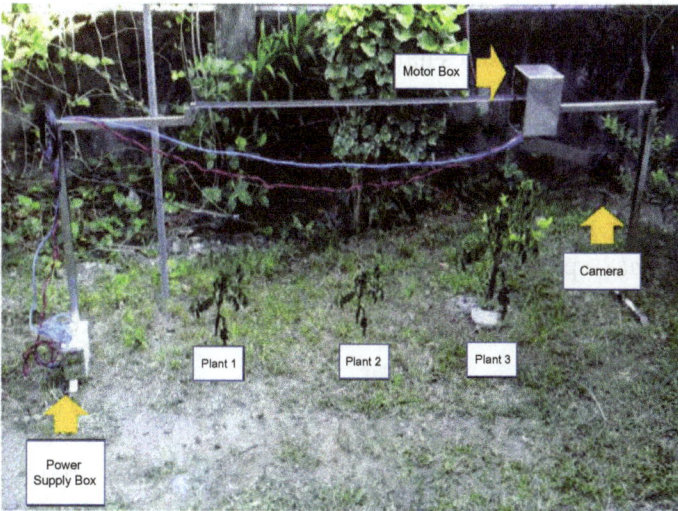

Fig. 3: Prototype of automatic image capturing box.

Note: Cited from De *et al.* Automated image capturing system for deep learning-based tomato plant leaf disease detection and recognition.

Fig. 4: Prediction of the proposed model on the test dataset and (a)–(d) are images that correctly detected as bacterial ones and (e)–(h) are correctly detected as healthy leaves.

Note: Cited from Yadav *et al.* Identification of disease using deep learning and evaluation of bacteriosis in peach leaf.

3.3. *Abiotic crop stress*

Global warming intensifies natural disasters, which seriously threaten the stable production of agriculture. Table 3 lists some representative works of deep learning in plant abiotic stress analysis.

Nitrogen, as the main component of amino acids, proteins, nucleic acids, and chlorophyll, is an essential nutrient for crop growth. Lack of nitrogen will affect crop height, color, and number of leaves, etc. Azimi et al.[37] proposed an image-based method for classifying the level of nitrogen deficiency stress in plants, which achieves accuracy of 75% when the picture has background. Without background, the accuracy is 84%.

Rapid determination of crop water stress is a prerequisite for timely and effective irrigation. Chandel et al.[38] identified crop water stress based on images, as shown in Fig. 5. They compared AlexNet, GoogLeNet, and Inception-V3. GoogLeNet got the best accuracy of 98.3%, 97.5%, and 94.1% on corn, okra, and soybean, respectively.

Finding suitable crops for saline land has long troubled local farmers. Feng et al.[39] used hyperspectral imaging to monitor the plant phenotypes of 13 okra (*Abelmoschus esculentus* L.) genotypes after 2 and 7 days of salt treatment. They found that different okra genes showed different levels of performance under salt stress, such as the level of fresh weight, SPAD, elemental content, and photosynthesis-related parameters. They introduced a plant leaf

Table 3: Application of deep learning in crop abiotic stress.

Application scenario	Author	Year	Algorithm
Measure stress level in plants due to nitrogen deficiency	Azimi et al.[37]	2021	CNN
Identifying crop water stress	Chandel et al.[38]	2021	DCNN
Obtain high-throughput plant salt	Feng et al.[39]	2020	CNN
Qualification of Soybean Responses to Flooding Stress	Zhou et al.[40]	2021	FNN
Abiotic stress (frost) in in-filed maize crop	Goswami et al.[41]	2019	RF
Application of crop drought stress	Sun et al.[42]	2019	SAE

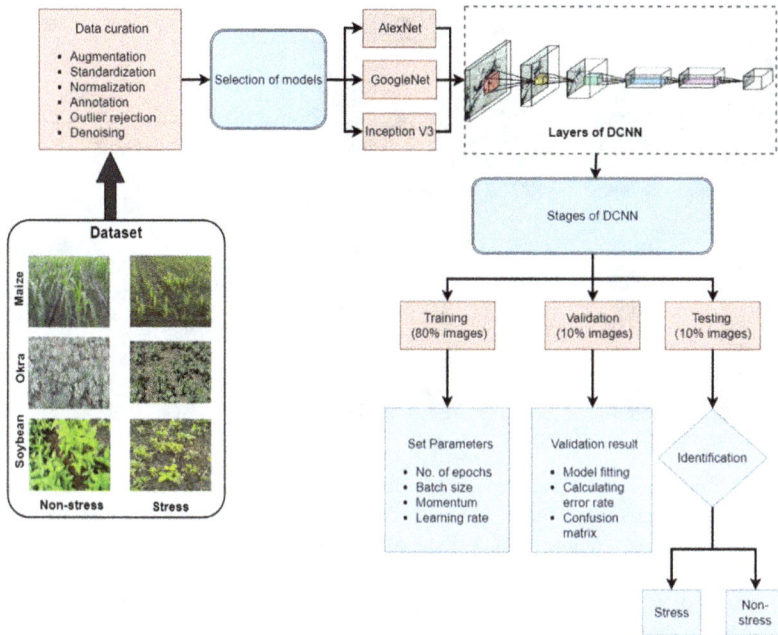

Fig. 5: Schematic diagram of the flow of work for identification of stressed crop.

Note: Cited from Chandel *et al.* Identifying crop water stress using deep learning models.

segmentation method in the hyperspectral image based on deep learning and constructed four sample prediction methods based on spectral data, with correlation coefficients of 0.835, 0.704, 0.609, and 0.588 for SPAD, sodium concentration, photosynthesis rate, and evapotranspiration rate, respectively. This result confirmed the effectiveness of high-throughput photographic technique in studying plant salt stress.

Soybean is the main crop in many areas. Flood stress affects the quality of soybean seeds and reduces soybean yield. Zhou *et al.*[40] proposed to screen for flood tolerance genes with crop images. They used drones to capture aerial images with five-band multispectral and infrared (IR) thermal cameras at altitudes of 20, 50, and 80 m off the ground (Fig. 6). Then they used an FNN for feature extraction and prediction and found that using images at 20 m had the best classification performance (0.9) in the level 5 flood damage score.

Frost damage can cause great loss of crop production. Goswami *et al.*[41] used multispectral drones to capture images of frost affected

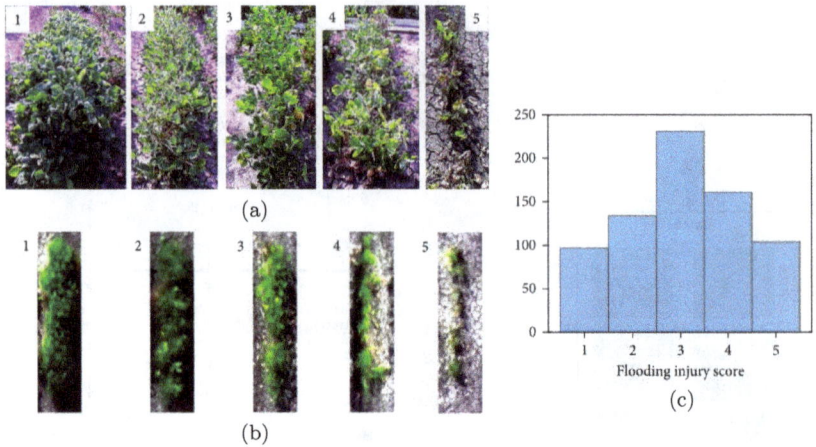

Fig. 6: Representative images of soybean plots of different flooding injury scores (FISs). (a) Images were taken using a consumer-grade camera to show example soybean plots at 1–5 levels of FISs. (b) Soybean plots of different FISs show differently in the UAV images composed of the red, green, and blue channels from the multispectral images. (c) Histogram of the visually observed FISs for 724 soybean plots.

Note: Cited from Zhou *et al.* Qualification of soybean responses to flooding stress using UAV-based imagery and deep learning.

maize in the field. They used these multispectral images to obtain changes in physiological structures and physiological parameters during plant growth. By comparing random forest (RF), random committee (RC), support vector machine (SVM), and neural network, they found that RF obtained a relatively high overall accuracy (86.47%) through kappa index analysis (KIA = 0.80). Their method was also quite cost-effective in terms of the computation time (0.08 s).

Developing crops that is resistant to drought can increase crop yields in arid and semi-arid areas, which is achieved through the screening of drought-resistant genotypes. Sun *et al.*[42] used chlorophyll fluorescence imaging to monitor the time series response of salt overly sensitive (SOS) mutants in drought conditions (Fig. 7). They extracted a fluorescent fingerprint map of the amoeba SOS mutants under drought stress. The chlorophyll fluorescence features are extracted by the SAE neural network to identify the amoeba chlorophyll fluorescent fingerprint map. They achieved 95% accuracy.

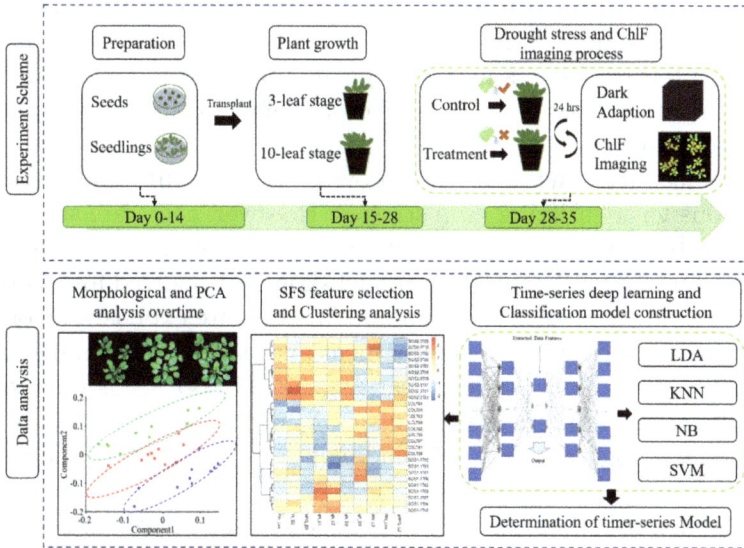

Fig. 7: Workflow of using chlorophyll fluorescence imaging to dynamically monitor photosynthetic fingerprints caused by some genes under drought conditions. LDA, linear discriminant analysis; KNN, k-nearest neighbor classifier; NB, Gaussian naive Bayes; SVM, support vector machine.

Note: Cited from Sun *et al.* Time-series chlorophyll fluorescence imaging reveals dynamic photosynthetic fingerprints of SOS mutants to drought stress.

4. Summary and trend

This chapter summarizes the latest developments of deep learning in crop stress. Overall, the application of deep learning in crop stress has the following advantages:

(1) It can improve the accuracy of classification, detection, and recognition. It overcomes the shortcomings of manual identification, which is prone to subjective influence, such as misjudgment and reliance on expert experience.
(2) Deep learning has good generalization capability.
(3) Compared with the traditional methods, it can increase the training set by data enhancement and expansion method.

Deep learning solves many previous unsolvable problems. Excellent research results have been achieved in both biotic stress and abiotic stress.

The effective identification and classification of crop stress can help scientists identify the relationship between genomic and phenotypes. In terms of stress monitoring, the problem of identifying crop stress in the early stage, where the crops show no obvious symptoms, still needs to be solved.

The deep learning algorithm provides good results under certain situations for identifying crop stresses. However, there are still many issues to be solved. First, the training cost of deep learning model is high and we do not know which model should be selected for a certain task. Second, the data collection and labeling are expensive. Considering the complex lighting environment and background changes, the generalization ability of deep learning model also needs our attention. Third, since there are various kinds of sensors, it is required to leverage heterogeneous data, including images, to improve the model performance.

In summary, deep learning has brought huge opportunities and new ideas to the research and application of crop stress.

References

1. A. C. Tyagi, Towards a second green revolution, *Irrigation and Drainage.* **4**(65), 388–389 (2016).
2. S. Lv, D. Li, and R. Xian, Research status of deep learning in agriculture of china, *Computer Engineering and Applications.* **55**(20), 24–33 (2019).
3. L. Saxena and L. Armstrong, A survey of image processing techniques for agriculture, *Proceedings of Asian Federation for Information Technology in Agriculture.* 401–413 (2014).
4. A. Singh, B. Ganapathysubramanian, A. K. Singh, and S. Sarkar, Machine learning for high-throughput stress phenotyping in Plants, *Trends in Plant Science.* **21**(2), 110–124 (2016).
5. K. He, X. Zhang, S. Ren, and J. Sun. Delving deep into rectifiers: Surpassing human-level performance on imagenet classification, in *Proceedings of the IEEE International Conference on Computer Vision*, 2015, pp. 1026–1034.
6. L. Dong, L. Su, and C. Zhi-dong, State-of-the-art on deep learning and its application in image object classification and detection, *Computer Science.* **43**(12), 13–23 (2016).
7. P. Viola, M. Jones, *et al.*, Robust real-time object detection, *International Journal of Computer Vision.* **4**(34–47), 4 (2001).

8. W. Wang, G. Wang, A. Bhatnagar, Y. Zhou, C. Xiong, and R. Socher, An investigation of phone-based subword units for end-to-end speech recognition, arXiv preprint arXiv:2004.04290 (2020).

9. Y. LeCun, Y. Bengio, and G. Hinton, Deep learning, *Nature.* **521** (7553), 436–444 (2015).

10. A. Krizhevsky, I. Sutskever, and G. E. Hinton, Imagenet classification with deep convolutional neural networks, *Advances in Neural Information Processing Systems.* **25**, 1106–1114 (2012).

11. J. Deng, W. Dong, R. Socher, L.-J. Li, K. Li, and L. Fei-Fei. Imagenet: A large-scale hierarchical image database, in *2009 IEEE Conference on Computer Vision and Pattern Recognition*, 2009, pp. 248–255.

12. W. Rawat and Z. Wang, Deep convolutional neural networks for image classification: A comprehensive review, *Neural Computation.* **29**(9), 2352–2449 (2017).

13. M. D. Zeiler and R. Fergus, Visualizing and understanding convolutional networks, in *European Conference on Computer Vision*, 2014, pp. 818–833.

14. K. He, X. Zhang, S. Ren, and J. Sun, Spatial pyramid pooling in deep convolutional networks for visual recognition, *IEEE Transactions on Pattern Analysis and Machine Intelligence.* **37**(9), 1904–1916 (2015).

15. K. Simonyan and A. Zisserman, Very deep convolutional networks for large-scale image recognition, arXiv preprint arXiv:1409.1556 (2014).

16. C. Szegedy, W. Liu, Y. Jia, P. Sermanet, S. Reed, D. Anguelov, D. Erhan, V. Vanhoucke, and A. Rabinovich, Going deeper with convolutions, in *Proceedings of the IEEE Conference on Computer Vision and Pattern Recognition*, 2015, pp. 1–9.

17. K. He, X. Zhang, S. Ren, and J. Sun, Deep residual learning for image recognition, in *Proceedings of the IEEE Conference on Computer Vision and Pattern Recognition*, 2016, pp. 770–778.

18. G. E. Hinton and R. R. Salakhutdinov, Reducing the dimensionality of data with neural networks, *Science.* **313**(5786), 504–507 (2006).

19. R. Girshick, J. Donahue, T. Darrell, and J. Malik, Rich feature hierarchies for accurate object detection and semantic segmentation, in *Proceedings of the IEEE Conference on Computer Vision and Pattern Recognition*, 2014, pp. 580–587.

20. P. Sermanet, D. Eigen, X. Zhang, M. Mathieu, R. Fergus, and Y. LeCun, Overfeat: Integrated recognition, localization and detection using convolutional networks, arXiv preprint arXiv:1312.6229 (2013).

21. R. Girshick, Fast R-CNN, in *Proceedings of the IEEE International Conference on Computer Vision*, 2015, pp. 1440–1448.

22. S. Ren, K. He, R. Girshick, and J. Sun, Faster R-CNN: Towards real-time object detection with region proposal networks, *Advances in Neural Information Processing Systems.* **28**, 91–99 (2015).
23. W. Liu, D. Anguelov, D. Erhan, C. Szegedy, S. Reed, C.-Y. Fu, and A. C. Berg, Ssd: Single shot multibox detector, in *European Conference on Computer Vision*, 2016, pp. 21–37.
24. J. Redmon, S. Divvala, R. Girshick, and A. Farhadi, You only look once: Unified, real-time object detection, in *Proceedings of the IEEE Conference on Computer Vision and Pattern Recognition*, 2016, pp. 779–788.
25. J. Redmon and A. Farhadi, YOLO9000: Better, faster, stronger, in *Proceedings of the IEEE Conference on Computer Vision and Pattern Recognition*, 2017, pp. 7263–7271.
26. J. Redmon and A. Farhadi, YOLOv3: An incremental improvement, arXiv preprint arXiv:1804.02767 (2018).
27. A. Bochkovskiy, C.-Y. Wang, and H.-Y. M. Liao, YOLOv4: Optimal speed and accuracy of object detection, arXiv preprint arXiv:2004. 10934 (2020).
28. A. K. Singh, B. Ganapathysubramanian, S. Sarkar, and A. Singh, Deep learning for plant stress phenotyping: Trends and future perspectives, *Trends in Plant Science.* **23**(10), 883–898 (2018).
29. J. Shin, Y. K. Chang, B. Heung, T. Nguyen-Quang, G. W. Price, and A. Al-Mallahi, A deep learning approach for RGB image-based powdery mildew disease detection on strawberry leaves, *Computers and Electronics in Agriculture.* **183**, 106042 (2021).
30. R. Ahila Priyadharshini, S. Arivazhagan, M. Arun, and A. Mirnalini, Maize leaf disease classification using deep convolutional neural networks, *Neural Computing and Applications.* **31**(12), 8887–8895 (2019).
31. R. G. De Luna, E. P. Dadios, and A. A. Bandala, Automated image capturing system for deep learning-based tomato plant leaf disease detection and recognition, in *TENCON 2018—2018 IEEE Region 10 Conference*, 2018, pp. 1414–1419.
32. S. Nickolas *et al.* Deep learning based betelvine leaf disease detection (piper betlet.), in *2020 IEEE 5th International Conference on Computing Communication and Automation (ICCCA)*, 2020, pp. 215–219.
33. J. M. Duarte-Carvajalino, D. F. Alzate, A. A. Ramirez, J. D. Santa-Sepulveda, A. E. Fajardo-Rojas, and M. Soto-Suárez, Evaluating late blight severity in potato crops using unmanned aerial vehicles and machine learning algorithms, *Remote Sensing.* **10**(10), 1513 (2018).
34. S. Yadav, N. Sengar, A. Singh, A. Singh, and M. K. Dutta, Identification of disease using deep learning and evaluation of bacteriosis in peach leaf, *Ecological Informatics.* **61**, 101247 (2021).

35. M. B. Tahir, M. A. Khan, K. Javed, S. Kadry, Y.-D. Zhang, T. Akram, and M. Nazir. Withdrawn: Recognition of apple leaf diseases using deep learning and variances-controlled features reduction, *Microprocessors and Microsystems.* 1–24 (2021).
36. M. Azadbakht, D. Ashourloo, H. Aghighi, S. Radiom, and A. Alimohammadi, Wheat leaf rust detection at canopy scale under different LAI levels using machine learning techniques, *Computers and Electronics in Agriculture.* **156**, 119–128 (2019).
37. S. Azimi, T. Kaur, and T. K. Gandhi, A deep learning approach to measure stress level in plants due to nitrogen deficiency, *Measurement.* **173**, 108650 (2021).
38. N. S. Chandel, S. K. Chakraborty, Y. A. Rajwade, K. Dubey, M. K. Tiwari, and D. Jat, Identifying crop water stress using deep learning models, *Neural Computing and Applications.* **33**(10), 5353–5367 (2021).
39. X. Feng, Y. Zhan, Q. Wang, X. Yang, C. Yu, H. Wang, Z. Tang, D. Jiang, C. Peng, and Y. He, Hyperspectral imaging combined with machine learning as a tool to obtain high-throughput plant salt-stress phenotyping, *The Plant Journal.* **101**(6), 1448–1461 (2020).
40. J. Zhou, H. Mou, J. Zhou, M. L. Ali, H. Ye, P. Chen, and H. T. Nguyen, Qualification of soybean responses to flooding stress using UAV-based imagery and deep learning, *Plant Phenomics.* **2021**, 1–13 (2021).
41. J. Goswami, V. Sharma, B. U. Chaudhury, and P. Raju, Rapid identification of abiotic stress (frost) in in-filed maize crop using uav remote sensing, *The International Archives of Photogrammetry, Remote Sensing and Spatial Information Sciences.* **42**, 467–471 (2019).
42. D. Sun, Y. Zhu, H. Xu, Y. He, and H. Cen, Time-series chlorophyll fluorescence imaging reveals dynamic photosynthetic fingerprints of sos mutants to drought stress, *Sensors.* **19**(12), 2649 (2019).

Part II
Signal Applications

https://doi.org/10.1142/9789811266911_0005

Chapter 5

A Mixed Pruning Method for Signal Modulation Recognition Based on Convolutional Neural Network

Shuncheng Gao, Xuzhang Gao, Jinchao Zhou,
Zhuangzhi Chen, Shilian Zheng, and Qi Xuan[*]

Institute of Cyberspace Security,
Zhejiang University of Technology,
Hangzhou, P.R. China

[*]*xuanqi@zjut.edu.cn*

Convolutional Neural Networks (CNN) are commonly used deep learning models in automatic modulation recognition (AMR) tasks. However, the success of CNNs is accompanied by a significant increase in the computation and parameter storage costs, and how to reduce the FLOPs and model size of CNNs has become a focus. Network pruning is a mainstream technique to reduce these overheads without hurting original accuracy too much. In this chapter, we propose a mixed pruning method which combines filter-level and layer-level pruning methods. Furthermore, we also define two indicators to quantitatively evaluate the comprehensive performance of the efficiency and quality of the pruned model. Finally, we carry out experiments on the public radio signal to demonstrate the effectiveness of our mixed pruning method.

1. Introduction

Since Tim O'Shea released the simulated signal dataset RML2016.
10a generated by GNU Radio software,[1] various neural network
models have also been applied to the signal modulation recognition
task.[2] Among them, the model based on convolutional neural network showed superior performance.

Although convolutional neural networks have been widely used
in artificial intelligence applications and have achieved great success, they still have shortcomings that cannot be ignored, that is,
high computational complexity and huge parameters.[3] When faced
with different learning tasks, the design of neural network models is
often crucial.[4,5] The traditional view of learning theory holds that a
good neural network model design should achieve a good trade-off
between the number of parameters and the expression capacity of
the model.[6,7] However, it is difficult to determine the appropriate
balance point in practice, so the pruning object is usually an overparameterized neural network model that has sufficient expressive
ability and is easy to optimize.[8–14] Although the greater the depth
and width of the neural network, the better the expression ability,
it also faces the problem of more resource consumption. For example, a 152-layer ResNet has more than 60 million parameters, and
the inferred resolution is 224×224 images require floating point
operations over 20 Gb.

This is just an experiment and an attempt at the scientific
research level. If we are considering transplanting the landing application to embedded or mobile devices, it will be subject to many
constraints such as significant memory usage, high computing capacity, and high energy consumption. So some technical methods have
been designed to alleviate the above problems.

The popular methods now include model lightweight design,
model weighting, low-rank decomposition, knowledge distillation,
and model pruning. Among these methods, neural network pruning has demonstrated its practicability, effectiveness, and efficiency
in various emerging applications. Pruning is a post-processing model
compression scheme.[15–17] On the premise of not losing useful information or not seriously affecting model performance, pruning removes
weak connections or less contributing neurons in neural networks to

obtain a smaller model with less performance loss and faster inference speed.

According to deleting the entire neuron or filter, pruning can be divided into unstructured pruning and structured pruning. Unstructured pruning considers each element in the weight of each filter, and deletes the weight parameter with value 0, thereby obtaining a sparse weight matrix.[18-20] Unstructured pruning has the highest flexibility and generalization performance. Generally, it can achieve a higher compression ratio. Through some regularization methods (L1 norm, Group-Lasso, hierarchical group-level sparse, etc.), the neural network model can adaptively adjust multiple granular structures during the training process (weight filter level, channel level, layer level, etc.) to achieve the effect of structured sparseness. Weight pruning[24] can be accelerated on dedicated software or hardware, but only on general hardware or BLAS libraries. It is difficult for the pruned model to achieve substantial performance improvement. The structured pruning method[21-23] is roughly divided into filter pruning, channel pruning, and layer pruning according to the granularity of pruning of the model structure. Since the entire filter and channel of some layers in the model are removed, the model structure is very regular. Not limited by hardware, it can significantly reduce the number of model parameters and obtain a significant acceleration effect. However, due to the coarse pruning granularity, it has a greater impact on the performance of the fine-tuned model, causing an irreparable loss of accuracy in the classification task. The core of structured pruning lies in the selection criteria of the model structure at different granularities, and the goal should be the lowest accuracy loss in exchange for the highest compression ratio.

To better achieve the goal, in this chapter we propose a mixed pruning method based on structured pruning, which combines filter-level and layer-level pruning methods to construct a simple and effective neural network pruning scheme. This method provides a new idea for solving scenarios where large-scale neural network models need to be deployed under limited resources in real life. Then we apply the mixed pruning method to the signal modulation recognition classification task. According to a given pruning ratio, unimportant filters or neural network layers are determined to be pruned. After pruning, fine-tuning is used to compensate for the loss of accuracy in our method. Compared with the original model, the final network is greatly reduced in terms of model size, running memory, and computing operation. At the same time, compared with a single method of pruning, our mixed pruning in terms of parameters and FLOPS can achieve a higher compression ratio and a more reasonable loss of model accuracy.

The rest of this chapter is organized as follows. In Section 2, we introduce the related work of neural network pruning in detail. In Section 3, we introduce our proposed mixed pruning method in more detail. In Section 4, we conduct experiments on the filter pruning methods, the layer pruning methods, and the mixed pruning methods, respectively, and further analyze the experimental results under different pruning methods. In Section 5, we briefly summarize our work.

2. Related Work

2.1. *Signal modulation recognition*

Automatic modulation recognition is a key technology in wireless communication. It is the basis of parameter estimation, spectrum monitoring, modulation and demodulation, and other communication technologies. It is widely used in many fields such as information interception and interference selection. However, with the development of communication technology, the wireless communication environment has become more and more complex, and the types of modulation have become more diverse. These have brought huge challenges to modulation recognition technology. It is imminent to study efficient and intelligent automatic modulation recognition methods. Deep learning, as a new technology in the field of artificial intelligence, has broken through many scientific and technological bottlenecks, and has been applied to many fields such as image recognition, text classification, and speech recognition. Therefore, utilizing deep learning technology to study signal automatic modulation recognition is the current and future development trend. As an end-to-end deep learning method, given the input signal, it can directly determine the modulation type of the given signal. To briefly summarize, there are three main ideas for signal modulation recognition based on deep learning at present:

(1) Based on one-dimensional convolutional neural network. O'Shea *et al.*[2] used one-dimensional convolutional neural network to directly perform feature extraction and recognition of signal sequences.
(2) Based on recurrent neural network, Rajendran *et al.*[25] used RNN (Recurrent Neural Network) and LSTM (Long Short-term Memory Network) to directly extract features and recognize time-domain signals.
(3) Based on two-dimensional convolution neural network,[1] the original modulated input signal was mapped into a two-dimensional input signal similar to an image. Then two-dimensional convolution was used to realize feature extraction and classification of the image, and make full use of the advantages of deep learning in the field of image research, i.e., quantities of neural network models and powerful performance.

2.2. *Convolutional neural network (ResNet56)*

In this chapter, we use the ResNet56 neural network model as our base model. Here is a brief introduction to some structural information of the model. The block formed by $F(x) + x$ is called a residual block. As shown in Fig. 1, a ResNet is formed by a series of multiple similar residual blocks. And a residual block has two paths, $F(x)$ and x. The $F(x)$ path is called the residual path, and the x path is the identity mapping, which is called "shortcut". The symbol $\oplus$ in Fig. 1 represents the element-wise addition. And the dimensions of each dimension of $F(x)$ and x that are involved in the operation are required to be the same.[10] The ResNet structure is very easy to modify and expand. By adjusting the number of channels in the block and the number of stacked blocks, the width and depth of the network can be easily adjusted, so that networks with different expression capabilities can be obtained without worrying too much about the network degradation. As long as the training data are sufficient, better performance can be obtained by gradually deepening the network. And the details of the main convolutional layer, fully connected layer,

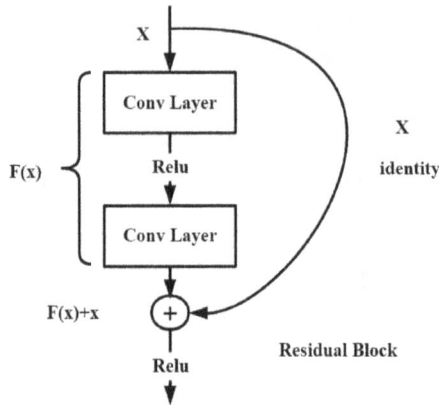

Fig. 1:   Residual block.

Table 1:   ResNet56 network model.

Layers	Output dimensions
Input	$(N, 1, 2, 128)$
Conv2d	$(N, 16, 2, 128)$
Residual stack $\times$ 9	$(N, 16, 2, 128)$
Residual stack $\times$ 9 Dowa-sampling module	$(N, 32, 1, 64)$
Residual stack $\times$ 9 Dowa-sampling module	$(N, 64, 1, 32)$
FC	$(N, 64)$
Softmax	$(N, 11)$

and classification layer of the ResNet56 neural network model used in this chapter are shown in Table 1.

In signal modulation classification task, the ResNet56 has an outstanding performance. Specifically, we map the original I/Q signal into a single-channel two-dimensional narrow image data form in the data processing stage.[1] Then we input the transformed data into the neural network model. Table 1 shows the transformation process of the data dimension flowing through the layers of the neural network model, where N represents the number of samples input to the neural network model for each batch. Finally, the probability of the category of each sample is output through the soft-max layer. In the typical pruning method, the difficulty of pruning the residual network model increases sharply due to the shortcut. But our proposed

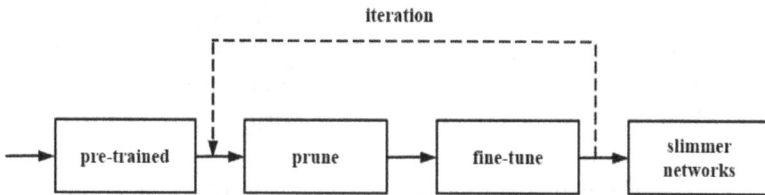

Fig. 2: The pruning process of ResNet56 network model.

mixed pruning algorithm can still perform well on this type of neural network model.

2.3. *Network pruning*

As shown in Fig. 2, the typical pruning process of a neural network includes three steps: (1) First, train an over-parameterized model; (2) Prune the trained large model according to a certain pruning standard, i.e., retain the important weight parameters or retain the important structure of the model; (3) Fine-tune the pruned neural network model to match the performance of the original model. In our proposed mixed pruning method, this typical pruning step is still used, and for convenience we denote the fine-tuned model by the pruned model. Of course, in different pruning strategies, iterative fine-tuning while pruning is also used.

2.3.1. *Low-rank decomposition*

Different from the pruning methods, the purpose of low-rank decomposition is to reduce the computational cost of the network, rather than focusing on changing the number of parameters of the neural network or the original number of filters. Low-rank decomposition utilized SVD (Singular Value Decomposition) and TD (Tucker Decomposition) decomposition techniques, and used low-rank matrices to approximate the weight matrix in the neural network.[26] Initially, this method is particularly effective on the fully connected layer of the neural network, and can compress the model to 1/3 of its original size. But it has no significant acceleration effect, since the calculation operations in CNN mainly come from the convolution operation of the convolution layer. To this end, Denton *et al.*[26] used the linear structure of CNN to approximate the convolution operation by exploring the low-rank approximation of weights and expressing the weight matrix as the low-rank product of two smaller matrices, and keeping the variation in the original model accuracy within 1%. Lin et al.[27] then proposed a low-rank decomposition of convolution filters and fully connected matrices. Although low-rank decomposition is conducive to the compression and acceleration of convolution neural networks, it usually leads to greater accuracy loss under high compression ratios. It is worth mentioning that the pruning technology of a neural network model is orthogonal to the low-rank decomposition technology. Therefore, we can apply low-rank decomposition on the basis of pruning to achieve a higher compression ratio and save more computing cost.

2.3.2. Weight pruning

The weight pruning method[28,29] first counts the contribution value of each weight parameter based on the Hessian matrix of the loss function, and then filters the weights with low contribution values. Usually, the neural network model is defined as a specific parameterized form $f(x; w)$, where w represents the specific neural network weight parameters obtained through the training dataset. The pruning operation of the neural network model needs to prune the "useless" weight parameters according to certain standards on the basis of the original pre-training model. After pruning, a new model will be generated, which can be expressed as $f(x; mw')$, where w' is a set of parameters different from w and $m \in \{0, 1\}^{w'}$ is a binary mask matrix that fixes certain weight parameters for convolutional or fully connected layers of the neural network model to 0.

Han et al.[20] proposed that the process of neural network training is not only learning weights, but also learning important connections of neural networks, and they prune the network weights based on the L1 norm criterion and retrain the network to compensate for the loss of accuracy. In the same year, they improved the pruning standard by combining quantization and Huffman coding for neural network compression.[19] Wen et al.[30] proposed to use Lasso regression during the model training to learn the optimal structure, including optimal shape of the filter, optimal number of channels, and optimal depth of neural network. Kang et al.[31] divided each filter into multiple groups and deleted the "useless" weights in each group. Denton et al.[26] point out that there is a smaller sparse network in the densely connected network, which can be trained from scratch to achieve the same accuracy as the dense network. And this idea opens up another new mode of neural network weight pruning.

2.3.3. *Filter and channel pruning*

In the filter and channel pruning methods, the pruning method[22] based on the L1 norm believes that the filter weight parameters with a smaller L1 norm have a small contribution to the final output of the neural network and can be removed safely. Among filter importance-based evaluation criteria, Molchanov *et al.*[32] evaluated the importance of a filter using a criterion based on the first-order Taylor expansion for the weights of neural networks, where the gradient and norm of the filter are used to approximate its global importance score. And He *et al.*[33] calculated the L1 norm of the filter weight parameters of each convolutional layer and used the geometric intermediate value to obtain the common information of the filter, considering pruning the redundant rather than important filter. The filter whose L1 norm is close to the geometric median value can be deleted safely. Luo *et al.*[34] proposed to determine the contribution of the filter to the output result according to the output value of each filter in the convolutional layer. Lin *et al.*[35] proposed to determine the importance of the filters according to the rank of the feature map generated by the convolutional layer, with the low-rank filters being cut off. Huang *et al.*[36] and Lin *et al.*[37] introduced new parameters (also known as mask matrices) to learn the sparse structure of neural networks, where the corresponding filters with zero values in the mask matrix are removed. He *et al.*[38] proposed to prune channels by channel selection based on Lasso regression and least squares reconstruction. Liu *et al.*[39] optimized the scaling factor γ of the BN layer as an indicator to measure the importance of the channel, and used the L1 norm to penalize the scaling factor for sparsity. Different from L1 norm regularization, Zhuang *et al.*[40] proposed a polarization regularization method which makes a clear distinction between the channels that need to be pruned and those that are reserved, and enhances the effect of pruning. Of course, there are some automatic channel pruning methods. For example, based on the artificial bee colony intelligent algorithm, Lin *et al.*[24] proposed to search for the best channel pruning ratio of each layer to determine the optimal model pruning structure. Some researchers[3] have also applied reinforcement learning to the pruning algorithms. Compared with traditional methods, automatic channel pruning can save a lot of manpower and may be more efficient.

2.3.4. *Layer pruning*

Different from the previous pruning methods, Chen *et al.*[41] proposed a layer pruning method based on feature representation. This method determines the redundant parameters by studying the features learned in the convolutional layer, and each layer is independently trained for layer sorting. After learning the importance of layers, the least important layers are pruned according to the learning results, simplifying the complex CNN into a more compact CNN with the same performance. Compared with the method of sorting after each layer is trained separately, Elkerdawy *et al.*[42] proposed a one-time layer sorting method of neural network models. The former does not have any improvement in the accuracy of model classification, while the latter proposes a set of layer pruning methods based on different pruning standards, which shows the limitations of the filter pruning method in reducing delay. Ro *et al.*[43] also advocate pruning at the layer level. After removing the layers that do not contribute to the model learning, automatically they proposed to set an appropriate learning rate for each layer. In this case, the model can improve performance during the fine-tuning phase.

3.   Methods

We have observed that in various pruning methods, when the pruning ratio reaches a certain level, the model will be in a relatively unhealthy state after pruning. In filter pruning, the width of the model is reduced. While in layer pruning, the depth of the model decreases. When the pruning rate rises to a certain level, the filter pruned model becomes narrow and long, while the layer pruned model becomes wider and flat. Like the stretching of a picture, if you only operate from one direction, it will lose its original information faster than operating from two directions. In this regard, can we prune the model in two directions at the same time, so that the pruned model has a healthier state, and its pruning effect is improved? To this end, we propose a mixed pruning method based on convolutional neural network for signal modulation recognition. Combining the two pruning methods, the model is pruned in both width and depth to ensure the model has good adaptability in width and depth. At the same time, it can further reduce the number of parameters and FLOPs of the neural network on the basis of less accuracy loss.

3.1. *Basic mixed pruning method framework*

The mixed pruning method comprehensively considers the filter pruning method and the layer pruning method, and its main structure is shown in Fig. 3. In the figure, we sort the filter importance and layer importance of the original model by a certain method according to the parameter values of the original model. On this basis, the original model will be pruned according to the pruning ratio, and a new model will be constructed. Finally, the new model will be fine-tuned.

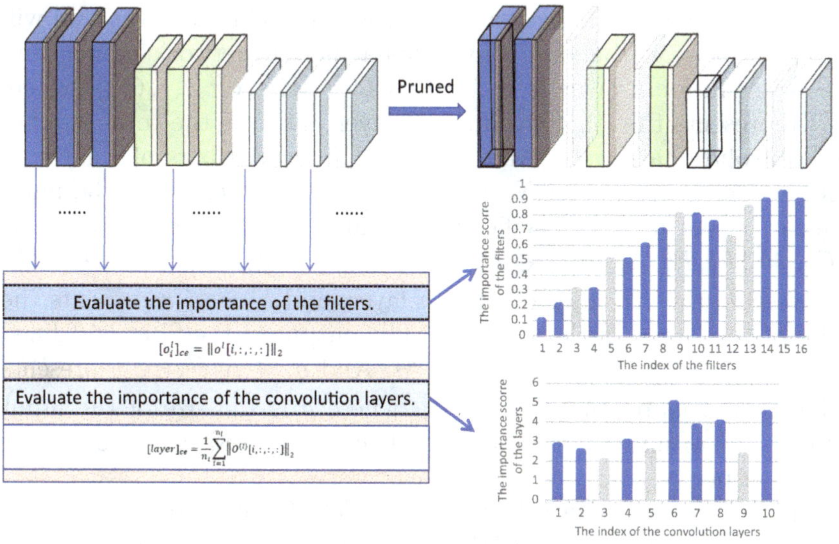

Fig. 3: The framework of mixed pruning.

Given an initial model, mixed pruning mainly includes the following steps:

(1) According to the parameters of the pre-trained model, qualitatively and quantitatively evaluate the importance of the filters in each layer with the selected parameter indicators, and rank them.
(2) According to the parameters of the pre-trained model, qualitatively and quantitatively evaluate the importance of each layer with the selected parameter indicators, and rank them.
(3) According to the given pruning ratio, determine the number of reserved layers and the number of filters in each layer, and define a new model.
(4) According to the importance indicators given in (1) and (2), copy the remaining layers and corresponding filter parameters from the original model to the new model.
(5) Train or fine-tune the new model, and the final model will be the pruned model.

3.2. Notations

In this section, we define some symbols and definitions that we will use. Suppose we build a convolutional neural network model, and its convolutional layer is defined as $C = \{C^1, C^2, \ldots, C^L\}$, where C^i represents the ith convolutional layer and L is the total number of convolutional layers. And the set of filter numbers of a layer in the model is defined as $N = (n_1, n_2, \ldots, n_L)$, where n^i is the number of filters in C^i. Specifically, the filters of C^i are expressed as $F_{C^i} = \{F_1^i, F_2^i, \ldots, F_{n_i}^i\} \in R^{n_i \times n_{i-1} \times k_i^1 \times k_i^2}$, where $F_j^i \in R^{n_{i-1} \times k_i^1 \times k_i^2}$ represents the jth filter of the ith layer, and $k_i^1 \times k_i^2$ represents the size of the kernel of the ith layer. The output of the filter is defined as $o^t = \{o_1^t, \ldots, o_{n_l}^t\} \in R^{n_l \times g \times h_l \times w_l}$. And $o_j^i \in R^{g \times h_i \times w_l}$ represents the feature map generated by F_j^t, where g is the size of the input data, and h_i and w_i represent the height and width of the feature map, respectively.

In this regard, if we define the model of the pre-trained neural network as M, the weight of the initial model after fine-tuning on the dataset as the model after pruning as M', and the weight of pruned model as W'. For the pruned model M', we denote its convolutional layer structure as $C' = (C^{\varphi_1}, C^{\varphi_2}, \ldots, C^{\varphi_K})$, where K is the number of layers of the model after pruning with $K \leq L$. And the filter structure is denoted as $F' = \left(f'_{\varphi_1}, f'_{\varphi_2}, \ldots, f'_{\varphi_K}\right)$, where φ_j represents the jth convolutional layer of the pruned model, f'_{φ_j} is the number of filters in the corresponding layer, and satisfies $C' \subseteq C$ and $f_{\varphi_j}{}' \leq f_{\varphi_g}$. Given the training set τ_{train} and the test set τ_{test}, for our mixed pruning, our goal is to find the best combination of C' and F', so that M' will have a higher compression ratio than M. At the same time, the pruned model M' can also achieve better accuracy than the pre-trained M.

To clarify our method, we have defined some new variables. Here, to indicate the relative accuracy of the pruned model, we define PR_{acc} as

$$\text{PR}_{\text{acc}} = \frac{\text{acc}(M'(C', F', W'; \tau_{\text{train}}); \tau_{\text{test}})}{\text{acc}(M(C, F, W; \tau_{\text{train}}); \tau_{\text{test}})}, \tag{1}$$

where $acc\left(M'\left(C', F', W'; \tau_{\text{train}}\right); \tau_{\text{test}}\right)$ means the accuracy of M' on the τ_{test}, and M' has a convolutional layer structure of C', a filter structure of F', with a weight of W' after training on the τ_{train}. To comprehensively evaluate the pruning rate of the model, we define PR_{FLOPs} and PR_{Param} as

$$PR_{\text{FLOPs}} = 1 - \frac{\text{FLOPS}\left(M'\left(C', F'\right)\right)}{\text{FLOPS}\left(M\left(C, F'\right)\right)}, \tag{2}$$

$$PR_{\text{Param}} = 1 - \frac{\text{Param}\left(M'\left(C', F'\right)\right)}{\text{Param}(M(C, F))}, \tag{3}$$

where $\text{FLOP}_{\text{s}}\left(M'\left(C', F'\right)\right)$ and $\text{Param}\left(M'\left(C', F'\right)\right)$ represent the FLOPs and parameter values of the model M' with the structure of C' and F'. So that the mixed pruning problem can be expressed as

$$M'\left(C', F'\right) = \text{argmax}\{A_1 \times \text{PR}_{\text{acc}} + A_2 \times \text{PR}_{\text{FLOP}_{\text{s}}} \\ + A_3 \times \text{PR}_{\text{Params}} + A_4\}, \tag{4}$$

where $A_{i=1,2,3,4}$ are proportional constants. It can be seen that our best pruning model is not fixed, we need to strike a balance between pruning structure and performance of the model.

3.3. Filter pruning part

In filter pruning part, we will quantify the importance of the filters of each layer according to a certain standard, and prune the filters of each convolutional layer according to the pruning rate. Taking the C^i layer as an example, we divide $F_{c^i} = \{F_1^i, F_2^i, \ldots, F_{n_i}^i\}$ into two parts, the important group is denoted as $I_{C^i} = \{F_{I_1^i}^i, F_{I_2^i}^i, \ldots, F_{I_{n_{i1}}^i}^i\}$, and the relatively unimportant group is denoted as $U_{C^i} = \{F_{U_1^i}^i, F_{U_2^i}^i, \ldots, F_{U_{n_{i2}}^i}^i\}$, where n_{i1} and n_{i2} represent the number of important filters and the number of unimportant filters, respectively. And these two parts satisfy the conditions that $n_i = n_{i1} + n_{i2}$, $I_{C^i} \cup U_{C^i} = F_{C^i}$, and $I_{C^i} \cap U_{C^i} = \emptyset$ at the same time.

Then the unimportant filters in each layer of the convolutional layer will be removed in filter pruning. For example, at the ith layer, we will remove U_{C^i}. At the same time, we will also remove the corresponding output feature map.

In this chapter, we use the feature map of each layer to quantify the importance of filters. For the lth convolutional layer, the feature map is $O^l = \{o_1^l, o_2^l, \ldots, o_{n_l}^l\} \in R^{m_l \times o \times n_l \times w_l}$. Then, we can calculate the statistical average of the weights of the feature map, which can be denoted as

$$\left[o_i^l\right]_{ce} = \left\|o^l[i, :, :, :]\right\|_2, \tag{5}$$

where $\left[o_i^l\right]_{ce}$ represents the quantitative value of filter importance, l represents the index of the layer, and i represents the filter index of the layer.

3.4. Layer pruning part

In layer pruning part, we will also quantify the importance of layers according to a certain standard. And then remove the convolutional layer according to the pruning rate. We divide all convolutional layers $C = \{C^1, C^2, \ldots, C^L\}$ into two parts, the important group is denoted as $I_C = \{C^{I_1}, C^{I_2}, \ldots, C^{I_{L_1}}\}$, and the relatively unimportant group is denoted as $U_C = \{C^{U_1}, C^{U_2}, \ldots, C^{U_{L_2}}\}$, where L_1 and L_2 represent the number of important filters and the number of unimportant filters, respectively. At the same time, these two parts satisfy the conditions that $L_1 = L_1 + L_2$, $I_C \cup U_C = C$, and $I_C \cap U_C = \emptyset$.

In the ResNet56 model, when we remove the unimportant layer U_C, we will also remove the residual block corresponding to the convolutional layer and connect the upper part and the next part of the residual block.

Similar to the ordering within layers of filter pruning, layer pruning is the ordering of importance among layers. In this chapter, we utilize the feature map of each layer to quantify the importance of the convolutional layer. For the lth convolutional layer, the feature map is $O^l = \{o_1^l, o_2^l, \ldots, o_{n_l}^l\} \in R^{n_l \times g \times h_l \times w_l}$. Then, we can calculate the statistical average of O^l, which can be denoted as

$$[\text{layer}]_{ce} = \frac{1}{n_l} \sum_{i=1}^{n_l} \left\| O^{(l)}[i, :, :, :] \right\|_2, \tag{6}$$

where $[\text{layer}]_{ce}$ represents the quantitative value of layer importance, l represents the index of the layer, and i represents the filter index of the layer.

4. Experiment

4.1. *Datasets*

The dataset used in the experiments is the public electromagnetic signal dataset RML2016.10a. It contains 11 modulation categories, i.e., 8PSK, AM-DSB, AM-SSB, BPSK, CPFSK, GFSK, PAM4, QAM16, QAM64, QPSK, WBFM. Each modulation category includes 20 kinds of signal-to-noise ratio (SNR) signals ranging from –20 dB to 18 dB in 2 dB steps. There are 1,100 signals per SNR for a total of 220,000 signals in the dataset.

In the experiments, a total of 11,000 signals with the highest signal-to-noise ratio of 18 dB in the original dataset are used. And the training set and test set are split as follows: 80% is used as the training set, with a total of 8,800 signals, and the remaining 2,200 signals are used as the test set.

4.2. *Baselines*

In this chapter, the proposed model is compared with the following models: (1) the original ResNet56 model, (2) the model after filter pruning based on the original model with the pruning rate of 0.2, 0.5, 0.7, (3) the model after layer pruning on the original model with the pruning rate of 0.2, 0.5, 0.7, (4) the model with the maximum pruning rate of each pruning method, in which the accuracy does not drop by more than 3%.

4.3. *Evaluation metrics*

In the pruning algorithm, we need to strike a reasonable balance between model efficiency and model quality. Generally speaking, the pruning algorithm usually increases the former, that is, improves the efficiency of the model while reducing the latter. This means that the best feature of the pruning method is not the single model it prunes, but a series of models corresponding to different points on the efficiency–quality curve.

In this study, we comprehensively use the average of the two commonly used indicators as a measure of the pruning effect. The first is the amount of multiplication and addition required for the network inference, denoted as FLOPs. The second is the number of parameters. Then we define an indicator PR_{xl} to quantify efficiency as follows:

$$PR_{xl} = \frac{PR_{FLOPs} + PR_{Param}}{2} * 100\%. \tag{7}$$

At the same time, in order to measure the quality of the pruned model, we define an indicator [Acc] as follows:

$$[Acc] = acc\left(M'\left(C', F', W'; \tau_{train}\right); \tau_{test}\right), \tag{8}$$

which is the accuracy of the pruned model in the classification task. Besides, in order to better evaluate the pruned model, we define an indicator PR_{zl} as follows:

$$PR_{zl} = \frac{PR_{FLOPs} + PR_{Param} + 98 \times PR_{acc}}{98}, \tag{9}$$

to describe the pros and cons of the pruned model on the efficiency–quality curve.

4.4. Results and analysis

In the experiments, we choose the improved ResNet56 model as the test model. This model is based on the basic ResNet56 model. By modifying the convolution kernel and some other parameters, the model has better adaptation to the signal data.

4.4.1. Results at the same pruning rate

We analyze the performance of mixed pruning on the 18 dB dataset. The amount of data is relatively small. We selected 80% as the training set and the rest as the test set. We compare filter pruning models, layer pruning models, and mixed pruning models with pruning rates of 0.2, 0.5, and 0.7, respectively, with the original model.

As shown in Table 2, we can see the performance of each pruning method on the dataset. Take the PR_{zl} of the initial model, which is 100, as the benchmark. When the pruning rate is 0.2, 0.5, and 0.7, the PR_{zl} values of filter pruned and mixed pruned models are all higher than that of the initial model, while the PR_{zl} values of layer pruned model are all lower than that of the initial model. It can be observed that when the pruning rate is 0.2, the accuracy of filter pruned model is 89.68%, which is 0.18% higher than the accuracy of the initial model. And its PR_{zl} values of 100.63 is the highest with a pruning ratio of 0.2. When the pruning rate rises to 0.5 and 0.7, mixed pruning has a better effect, and the PR_{zl} values of the

Table 2: Results of each pruning method at different pruning rates.

Model	[Acc](%)	FLOPs	Parameters	$PR_{xl}(\%)$	PR_{zl}
Original	89.50	$12.33M$	$287.50K$	/	100
Filter	89.68	$9.71M$	$228.17K$	20.93	100.63
	88.86	$6.23M$	$145.18K$	49.49	100.3
	88.55	$3.61M$	$85.85K$	70.42	100.37
Layer	88.73	$9.10M$	$243.47K$	20.74	99.56
	87.31	$5.89M$	$174.86K$	53.24	98.65
	87.77	$3.51M$	$100.36K$	68.32	99.47
Mixed	89.45	$9.40M$	$225.76K$	22.6	100.41
	89.05	$5.65M$	$149.02K$	51.17	100.54
	89.09	$3.27M$	$92.51K$	70.67	100.99

Table 3:   Result of high pruning rate.

Model	[Acc](%)	FLOPs	Parameters	$PR_{xl}(\%)$	PR_{zl}
Original	89.50	12.33M	287.50K	/	100
Filter	87.41	2.42M	56.23K	80.42	99.3
Layer	87.63	1.92M	50.70K	83.4	99.62
Mixed	87.95	0.39M	10.48K	96.6	100.24

pruned model are 100.54 and 100.99, respectively, which are higher than those of filter pruned and layer pruned models.

4.4.2.   Result of high pruning rate

We also test the maximum pruning rate of each pruning method while guaranteeing effective pruning. We record the various indicators of each pruning method, with an accuracy loss of no more than 3%.

As shown in Table 3, we can see the performance of each pruning method on the dataset. The PR_{zl} value of our mixed pruning method is higher than 100, which means the pruning effect is the best. In addition, the PR_{xl} value of our mixed pruned model is 96.60%, which is much higher than 80.42% for the filter pruned model and 83.40% for the layer pruned model. At the same time, the FLOPs of the mixed pruned model are reduced by 96.85%, and the number of parameters is reduced by 96.36%.

4.4.3.   Analysis

Our mixed pruning has a relatively better performance on the 18 dB signal dataset. In terms of accuracy, our mixed pruning can be adjusted in two directions, making the model structure after mixed pruning more diverse. In terms of compression rate, our hybrid pruning comprehensively considers the filter level and the layer level, which can achieve higher compression rate than single-level pruning.

5.   Conclusion

In this research, we propose a mixed pruning method for signal modulation recognition based on convolutional neural network. We fuse the methods of filter pruning and layer pruning, and show more comprehensive capacity compared with adopting only a one-level pruning method. According to experimental verification, at the same pruning rate, our mixed pruning has a better and more stable balance of efficiency and quality than traditional pruning methods. In terms of compression rate, our mixed pruning can achieve a higher pruning rate with a smaller loss of precision. In the future, we will try to design a pruning algorithm that can adaptively adjust the balance between efficiency and effect according to tasks to meet the deployment requirements of edge devices.

References

1. T. J. O'Shea, J. Corgan, and T. C. Clancy. Convolutional radio modulation recognition networks, in *International conference on Engineering Applications of Neural Networks*, 2016, pp. 213–226.
2. T. J. O'Shea, T. Roy, and T. C. Clancy, Over-the-air deep learning based radio signal classification, *IEEE Journal of Selected Topics in Signal Processing.* **12**(1), 168–179 (2018).
3. Y. He, J. Lin, Z. Liu, H. Wang, L.-J. Li, and S. Han. AMC: AutoML for model compression and acceleration on mobile devices, in *Proceedings of the European Conference on Computer Vision (ECCV)*, 2018, pp. 784–800.
4. Y. LeCun, Y. Bengio, and G. Hinton, Deep learning, *Nature.* **521** (7553), 436–444 (2015).
5. J. Schmidhuber, Deep learning in neural networks: An overview, *Neural Networks.* **61**, 85–117 (2015).
6. M. J. Kearns and U. Vazirani, *An Introduction to Computational Learning Theory.* Cambridge, UK: The MIT Press, 1994.
7. V. Vapnik, *The Nature of Statistical Learning Theory*, 1999, New York: Springer Science & Business Media.
8. A. Krizhevsky, G. Hinton, *et al.*, Learning multiple layers of features from tiny images, *Handbook of Systemic Autoimmune Diseases.* **1**, 1–60 (2009).
9. K. Simonyan and A. Zisserman, Very deep convolutional networks for large-scale image recognition, arXiv preprint arXiv:1409.1556 (2014).
10. K. He, X. Zhang, S. Ren, and J. Sun, Deep residual learning for image recognition. *Computer Vision and Pattern Recognition*, 2016, arXiv preprint arXiv:1512.03385 (2016).
11. G. Huang, Z. Liu, L. Van Der Maaten, and K. Q. Weinberger. Densely connected convolutional networks, in *Proceedings of the*

IEEE Conference on Computer Vision and Pattern Recognition, 2017, pp. 4700–4708.

12. M. Soltanolkotabi, A. Javanmard, and J. D. Lee, Theoretical insights into the optimization landscape of over-parameterized shallow neural networks, *IEEE Transactions on Information Theory.* **65**(2), 742–769 (2018).

13. D. Zou, Y. Cao, D. Zhou, and Q. Gu, Gradient descent optimizes over-parameterized deep ReLU networks, *Machine Learning.* **109**(3), 467–492 (2020).

14. Z. Allen-Zhu, Y. Li, and Y. Liang, Learning and generalization in over-parameterized neural networks, going beyond two layers, *Advances in Neural Information Processing Systems.* **32**, 6158–6169 (2019).

15. M. C. Mozer and P. Smolensky, Skeletonization: A technique for trimming the fat from a network via relevance assessment, *Advances in Neural Information Processing Systems.* **1**, 107–115 (1988).

16. E. D. Karnin, A simple procedure for pruning back-propagation trained neural networks, *IEEE Transactions on Neural Networks.* **1**(2), 239–242 (1990).

17. S. A. Janowsky, Pruning versus clipping in neural networks, *Physical Review A.* **39**(12), 6600 (1989).

18. M. A. Carreira-Perpinán and Y. Idelbayev. "learning-compression" algorithms for neural net pruning, in *Proceedings of the IEEE Conference on Computer Vision and Pattern Recognition*, 2018, pp. 8532–8541.

19. S. Han, H. Mao, and W. J. Dally, Deep compression: Compressing deep neural networks with pruning, trained quantization and Huffman coding, arXiv preprint arXiv:1510.00149 (2015).

20. S. Han, J. Pool, J. Tran, and W. Dally, Learning both weights and connections for efficient neural network, *Advances in Neural Information Processing Systems.* **28**, 1135–1143 (2015).

21. J. Park, S. Li, W. Wen, P. T. P. Tang, H. Li, Y. Chen, and P. Dubey, Faster CNNs with direct sparse convolutions and guided pruning, arXiv preprint arXiv:1608.01409 (2016).

22. H. Li, A. Kadav, I. Durdanovic, H. Samet, and H. P. Graf, Pruning filters for efficient convNets, arXiv preprint arXiv:1608.08710 (2016).

23. M. Lin, L. Cao, S. Li, Q. Ye, Y. Tian, J. Liu, Q. Tian, and R. Ji, Filter sketch for network pruning, *IEEE Transactions on Neural Networks and Learning Systems.* **33**(12), 7091–7100 (2022).

24. M. Lin, R. Ji, Y. Zhang, B. Zhang, Y. Wu, and Y. Tian, Channel pruning via automatic structure search, arXiv preprint arXiv:2001.08565 (2020).

25. S. Rajendran, W. Meert, D. Giustiniano, V. Lenders, and S. Pollin, Deep learning models for wireless signal classification with distributed low-cost spectrum sensors, *IEEE Transactions on Cognitive Communications and Networking.* 4(3), 433–445 (2018).
26. E. L. Denton, W. Zaremba, J. Bruna, Y. Lecun, and R. Fergus, Exploiting linear structure within convolutional networks for efficient evaluation, *Advances in Neural Information Processing Systems.* **27**, 1269–1277 (2014).
27. S. Lin, R. Ji, C. Chen, D. Tao, and J. Luo, Holistic CNN compression via low-rank decomposition with knowledge transfer, *IEEE. Transactions on Pattern Analysis and Machine Intelligence.* **41**(12), 2889–2905 (2018).
28. Y. LeCun, J. Denker, and S. Solla, Optimal brain damage, *Advances in Neural Information Processing Systems.* **2**, 598–605 (1989).
29. B. Hassibi and D. Stork, Second order derivatives for network pruning: Optimal brain surgeon, *Advances in Neural Information Processing Systems.* **5**, 164–171 (1992).
30. W. Wen, C. Wu, Y. Wang, Y. Chen, and H. Li, Learning structured sparsity in deep neural networks, *Advances in Neural Information Processing Systems.* **29**, 2082–2090 (2016).
31. H.-J. Kang, Accelerator-aware pruning for convolutional neural networks, *IEEE Transactions on Circuits and Systems for Video Technology.* **30**(7), 2093–2103 (2019).
32. P. Molchanov, S. Tyree, T. Karras, T. Aila, and J. Kautz, Pruning convolutional neural networks for resource efficient inference, arXiv preprint arXiv:1611.06440 (2016).
33. Y. He, P. Liu, Z. Wang, Z. Hu, and Y. Yang. Filter pruning via geometric median for deep convolutional neural networks acceleration, in *Proceedings of the IEEE/CVF Conference on Computer Vision and Pattern Recognition*, 2019, pp. 4340–4349.
34. J.-H. Luo, J. Wu, and W. Lin. Thinet: A filter level pruning method for deep neural network compression, in *Proceedings of the IEEE International Conference on Computer Vision*, 2017, pp. 5058–5066.
35. M. Lin, R. Ji, Y. Wang, Y. Zhang, B. Zhang, Y. Tian, and L. Shao. Hrank: Filter pruning using high-rank feature map, in *Proceedings of the IEEE/CVF Conference on Computer vision and Pattern Recognition*, 2020, pp. 1529–1538.
36. Z. Huang and N. Wang. Data-driven sparse structure selection for deep neural networks, in *Proceedings of the European Conference on Computer Vision (ECCV)*, 2018, pp. 304–320.

37. S. Lin, R. Ji, C. Yan, B. Zhang, L. Cao, Q. Ye, F. Huang, and D. Doermann. Towards optimal structured CNN pruning via generative adversarial learning, in *Proceedings of the IEEE/CVF Conference on Computer Vision and Pattern Recognition*, 2019, pp. 2790–2799.

38. Y. He, X. Zhang, and J. Sun. Channel pruning for accelerating very deep neural networks, in *Proceedings of the IEEE International Conference on Computer Vision*, 2017, pp. 1389–1397.

39. Z. Liu, J. Li, Z. Shen, G. Huang, S. Yan, and C. Zhang. Learning efficient convolutional networks through network slimming. in *Proceedings of the IEEE International Conference on Computer Vision*, 2017, pp. 2736–2744.

40. T. Zhuang, Z. Zhang, Y. Huang, X. Zeng, K. Shuang, and X. Li, Neuron-level structured pruning using polarization regularizer, *Advances in Neural Information Processing Systems.* **33**, 9865–9877 (2020).

41. S. Chen and Q. Zhao, Shallowing deep networks: Layer-wise pruning based on feature representations, *IEEE Transactions on Pattern Analysis and Machine Intelligence.* 41(12), 3048–3056 (2018).

42. S. Elkerdawy, M. Elhoushi, A. Singh, H. Zhang, and N. Ray, To filter prune, or to layer prune, that is the question, in *Proceedings of the Asian Conference on Computer Vision*, 2020, pp. 737–753.

43. Y. Ro and J. Y. Choi AutoLR: Layer-wise pruning and auto-tuning of learning rates in fine-tuning of deep networks, in *Proceedings of the AAAI Conference on Artificial Intelligence*, 2021, vol. 35, pp. 2486–2494.

Chapter 6

Broad Learning System Based on Gramian Angular Field for Time Series Classification

Tingting Feng, Zhuangzhi Chen, Dongwei Xu, and Qi Xuan*

Institute of Cyberspace Security,
Zhejiang University of Technology,
Hangzhou, P.R. China

*xuanqi@zjut.edu.cn

As a common form of data expression, time series has a wide range of applications and has high research value. In the context of today's big data era, time-series data are increasingly characterized by information redundancy, long data, and high dimensionality. The classification of time series has become an important and challenging task in data analysis. In recent years, many effective deep learning algorithms have been proposed for time series classification problems. Deep learning is very powerful in discovering complex structures in high-dimensional data, but there are problems with high computational costs as well as gradient explosions. To avoid these difficulties, we develop a novel method (namely, GAF–BLS) to classify time series data. With this method, the time features and deep abstract features of the time series are extracted through the Gramian Angular Field (GAF), and then in the broad learning system (BLS), these features are mapped to a more discriminative space to be further enhanced and complete the classification task. Experiments show that the GAF–BLS can significantly improve the recognition efficiency on the basis that the recognition rate is not reduced.

1. Introduction

Time series data are a collection of observations that are observed
in chronological order. Classification and prediction are classic problems in the field of data mining. However, due to the complexity
of time series data, in past research, time series data classification
has become a special challenge in classification research. A lot of
work is devoted to the development and improvement of time series
forecasting, including from univariate to multivariate, from offline to
online, from linear to nonlinear, from short time series to long time
series, etc.

However, although using the original form of time series as the
input of the classifier can preserve the nonlinear characteristics to
a certain extent, it does not fully consider its time correlation. In
addition, the training of the LSTM for processing time series has
very high hardware requirements, especially when processing a long
time series, a large amount of storage bandwidth and computing
resources are required to complete the training, thus it is also time-
consuming. At present, in the research of time series classification,
researchers have begun to transform one-dimensional time series into
two-dimensional data representation for the training and recognition
of classification models. Wang et al.[1] encode time series data into different types of images: Gramian Angular Field (GAF) and Markov
Transition Fields (MTF), to realize the classification of time series
data using the recognition models in the image field. Hatami et al.[2]
use Recurrence Plot to convert a one-dimensional time series into a
two-dimensional texture map for image recognition tasks. The above
methods all perform well in the classification of time series datasets.

The rapid development of neural networks provides convenient
conditions for data processing. Early neural networks mainly focused
on solving the problems of adjusting parameters and network
hierarchical structure, especially the solution of gradient descent
parameters, but there are inherent shortcomings , such as slow learning convergence speed, easy to fall into local minima. At present,
neural networks are widely used in many fields such as speech recognition, image processing, and object recognition,[3] and have achieved
great success. The most common neural network is the Back Propagation (BP) Network,[4] and recently deep learning models[5] have made a

big breakthrough in image processing, such as Convolutional Neural Networks (CNN),[6,7] Deep Belief Networks (DBN),[8] Deep Boltzmann Machine (DBM).[9] However, the DL networks have multilayer neural networks with large size and complex structure, so there is a problem of slow training speed. While the BP network with a relatively simple network structure also has the problem of slow iterative solution speed and is easy to fall into a local minimum solution.

To solve the bottleneck of the deep neural networks mentioned earlier, Chen *et al.*[10,11] propose a broad learning system (BLS) and prove its approximation ability.[12] The structure of BLS is very different from that of deep neural networks. Compared with the "depth" structure, BLS is more inclined to structure the network in the direction of "width". BLS processes the input data through feature mapping to generate feature nodes and then performs nonlinear transformations on the feature nodes to generate enhanced nodes. The feature nodes and enhanced nodes are spliced together as a hidden layer. The output of the hidden layer is connected to the weight to get the final output. BLS only needs to use the pseudo-inverse to calculate the weight from the hidden layer to the output layer of the network. This solution method is very fast and will not encounter the problem of gradient disappearance or explosion. In addition, when the test accuracy of the model does not meet the expected requirements, BLS can also use incremental learning methods to quickly achieve model reconstruction. After the emergence of BLS, BLS-based variants have been proposed to improve the performance of the method. These algorithms are widely used in image classification, pattern recognition, numerical regression, EEG signal processing, automatic control, and other fields.

Inspired by the literature,[1] this chapter proposes a classification method combining GAF and BLS. GAF is introduced to encode univariate time series, fully retaining the time dependence of the original time series. The feature matrix obtained after encoding is used as the input of feature nodes in the broad learning system, the feature node-set is generated after mapping as the input of the BLS enhancement node, and the enhancement layer output is generated through the mapping. The combination of the node-set and the enhancement layer is used to obtain the final output; and finally, the trained model is used to complete the classification task. The experimental results prove that the algorithm proposed in this chapter can extract feature information more effectively, as well as effectively improve the accuracy and efficiency of time series data recognition.

2. Preliminary

This chapter combines GAF and BLS to classify time series. Therefore, this part introduces the prior knowledge of these two parts, mainly including feedforward neural network, sparse autoencoder, and Gram Matrix.

2.1. *Feedforward neural network*

The section first takes single-layer feedforward neural networks (SLFN) as an example to introduce the model structure and mathematical expression of neural networks in two-class and multi-class problems. Then, the network training method based on gradient-descent learning algorithms is introduced. Finally, the expansion of the neural network in the structure is summarized.

2.1.1. *Single-layer feedforward neural networks*

The goal of classification is to classify the input variable x into one of K discrete categories C_k. The classification situation discussed in the chapter is that the categories do not intersect each other, that is, each input is classified into a unique category. The input space is divided into different decision-making regions, and its boundaries are called decision boundaries or decision surfaces.[13] This chapter models the conditional probability distribution based on the Bayesian neural network and uses the probability distribution $p(C_k|x)$ to make classification decisions.

A neural network is composed of nonlinear fundament functions, and each basis function is a nonlinear fundament function of the input variables. Suppose that $\{(x_1, y_1), ..., (x_n, y_n)\}$ is the data sample of the input layer, and the input data dimension is d, the number of hidden layer neuron nodes is M, and the number of output layer nodes is K, then a single hidden layer fully connected neural network structure can be constructed as shown in Fig. 1.

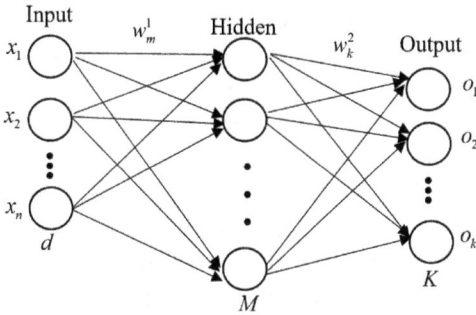

Fig. 1: Single hidden layer neural network.

The input $u_k^{(2)}$ and output $O_k^{(2)}$ of the kth node of the output layer are shown in Eqs. (1) and (2), respectively, where $k = 1, \ldots, K$:

$$u_k^{(2)}(x; w) = \sum_{j=1}^{M} \left(w_{kj}^{(2)} h \left(\sum_{i=1}^{n} (w_{ji}^{(1)} x_i + b_j^{(1)}) \right) + b_k^{(2)} \right) \qquad (1)$$

$$O_k^{(2)}(x; w) = \sigma(u_k(x; w)), \qquad (2)$$

where $w_{ji}^{(1)}$ represents the weight of connecting the ith neuron of the input layer to the jth neuron in the hidden layer, $b_j^{(1)}$ represents the bias of the jth neuron, and the superscript (1) indicates that the parameter belongs to the first layer of the neural network. $h(\cdot)$, $\sigma(\cdot)$ are the nonlinear activation functions of the hidden layer and the output layer, respectively. $h(\cdot)$ usually adopts a sigmoid function, such as a hyperbolic tangent function. $\sigma(\cdot)$ usually uses the Logistic Sigmoid activation function in binary classification problems. The Sigmoid function maps the input variables to $(0, 1)$ interval, which can be used to represent conditional probability $p(C_k|x)$.[13] In the case of binary classification:

$$p(C_k|x) = \text{sigmoid}(u_k) = \frac{1}{1 + \exp(-u_k)}. \qquad (3)$$

Equation (3) can be extended to the multi-class case ($K > 2$) through the normalization function:

$$p(C_k|x) = \frac{\exp(u_k)}{\sum_j^K \exp(u_j)}. \qquad (4)$$

Equation (4) is also called the softmax activation function.

The universal approximation theory shows that when the hidden layer has enough neurons, there is at least one layer of feedforward neural network with "squeezing" property activation function and the linear output layer can be approximated by any Borel measurable function mapped from one finite space to another finite space with arbitrary precision.[13] The "squeezing" property refers to squeezing the input value range to a certain output range.

Then we discuss the method of determining network parameters through maximum likelihood estimation in the binary classification problems. First, use the negative log-likelihood function as the error function to obtain the cross-entropy loss function:

$$
\begin{aligned}
E(w) &= -\ln L(x, y) \\
&= -\ln P(O = y|x; w) \\
&= -\sum_{n=1}^{N} y_n \ln O_n + (1 - y_n) \ln(1 - O_n).
\end{aligned}
\tag{5}
$$

In Eq. (5), $y_n \in \{0, 1\}$, $y_n = 1$ indicates that the category of the nth training sample is C_1, and $y_n = 0$ indicates that the category is C_2. O_n is the network output of the nth sample. When a sigmoid is used as the activation function, it satisfies $0 \le O_n \le 1$. Then, find the network parameter w that makes the selected error function $E(w)$ reach the minimum. Since the error function is a smooth continuous function, its minimum value appears in the parameter space where the gradient of the error function is equal to zero,[13] namely: $\nabla E(w) = 0$. In practical applications, the relationship between the error function and the network parameters is usually nonlinear, so for any local minimum point w, there are other equivalent minimum solutions in the parameter space. Since it is difficult to find an analytical solution to the equation, iterative numerical methods can be used to solve the continuous nonlinear function. Among the algorithms for training neural networks, the gradient descent algorithm based on the error function gradient information is widely used, and its parameter update method is shown in Eq. (6) as

$$
w^{(\tau+1)} = w^\tau - \eta \nabla E(w^\tau).
\tag{6}
$$

In the formula, the superscript τ represents the number of iterations, and the parameter $\eta > 0$ represents the learning rate. The gradient descent method mainly has the following three forms: batch gradient descent, stochastic gradient descent (SGD), and mini-batch gradient descent. The batch gradient descent algorithm uses all samples for parameter updates in each iteration. This method can better represent the sample population by using all the datasets, so it can more accurately face the direction of the extreme value of the objective function. But when the number of samples is large, the training process is slower. The SGD algorithm uses only one sample to update the parameters in each iteration, which can increase the training speed. But even when the objective function is strongly convex, SGD still cannot achieve linear convergence and may converge to a local optimum.[14] The mini-batch gradient descent algorithm uses appropriate batch size samples to update the parameters in each iteration, which is a compromise between the first two methods. Taking the SGD algorithm as an example, its parameter update mathematical expressions are shown as Eqs. (7) and (8), as follows:

$$E(w) = \sum_{n=1}^{N} E_n(w), \tag{7}$$

$$w^{\tau+1} = w^{\tau} - \eta \nabla E(w^{\tau}). \tag{8}$$

The update of formula (7) is repeated on the dataset cyclically, which can either process batch data sequentially, or randomly select data points repeatedly.

2.1.2. *RVFL neural network*

The neural network can be expanded by introducing a cross-layer link (direct connection), and a random weight neural network with a direct connection structure can improve the classification performance of the model. Random vector functional link (RVFL) is a type of neural network with random weight (NNRW). Compared with NNRW, RVFL increases the direct link from the input layer to the output layer. The network structure is shown in Fig. 2. Suppose

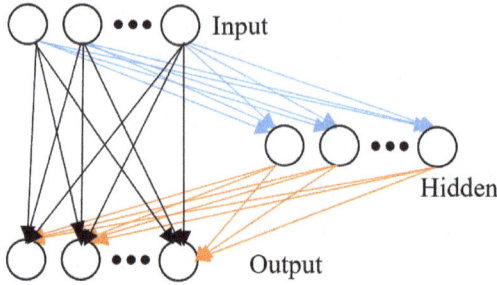

Fig. 2:   Random vector functional link network.

the network input is a d dimensional vector, $x \in R^d$ and the output $y \in R$. The RVFL neural network can be described as the weighted summation of the output of L hidden layer nodes, expressed as

$$f(x) = \sum_{m=1}^{L} \beta_m h_m(x; w_m) = \beta^T h(x; w_1, \ldots, w_L). \qquad (9)$$

In the formula, the input x is mapped and transformed by the random weight vector w_m to obtain the mth transformation. The $h : x \to R$ of each hidden node is called a basis, or function link, which usually uses the sigmoid function. The parameter w_m of the basis function is randomly selected from a pre-defined probability distribution and remains unchanged throughout the training process. Assuming that the network can provide enough hidden layer basis functions, and each basis function is continuous and stable, when L is large enough, RVFL has the ability of general approximation.[15] Suppose there are N samples $S = x_i, y_i, i = 1, ..., N$. There are usually two RVFL output parameter training algorithms: one is iterative, which iteratively obtains output weights based on the gradient of the error function. The other is to obtain output weights based on least squares.[15] Now use the regularized least square method to obtain the optimal parameter β as follows:

$$\beta^* = \arg \min_{\beta \in R^L} J(\beta) = \arg \min_{\beta \in R^L} \left(\frac{1}{2} \|H\beta - Y\|_2^2 + \frac{\lambda}{2} \|\beta\|_2^2 \right). \qquad (10)$$

In Eq. (10), the formula H is as follows:

$$H = \begin{bmatrix} h_1(x_1) & \cdots & h_1(x_L) \\ \vdots & \ddots & \vdots \\ h_N(x_1) & \cdots & h_N(x_L) \end{bmatrix}. \tag{11}$$

Let the gradient of $J(\beta)$ be zero as follows:

$$\frac{\partial J}{\partial \beta} = H^T H \beta - H^T Y + \lambda \beta = 0,$$

$$\beta^* = (HH^T + \lambda I)^{-1} H^T Y, \tag{12}$$

where I represents the identity matrix. In the case of $N \ll L$, it can be simplified by the following formula (13) to reduce the computational complexity of the inversion. For any $\lambda > 0$, there are

$$(H + \lambda I)^{-1} H^T = H^T (HH^T + \lambda I)^{-1}. \tag{13}$$

From Eqs. (12) and (13), another equation for solving the optimal weight β can be obtained:

$$\beta^* = H^T (HH^T + \lambda I)^{-1} Y. \tag{14}$$

Regularization of randomized weights can make RVFL converge quickly and alleviate the over-fitting problems.

2.2. *Sparse auto-encoder*

In supervised learning tasks, such as classification, to train a model with good classification performance, it is necessary to use good data features as a system input. Data features contain a lot of data information that is helpful for classification. Such features are usually extracted using very complex mathematical models. In some models, a set of random values can also be used to replace them. However, randomness is unpredictable and has a greater impact on classification performance. To reduce the impact of this kind of randomness on classification performance, sparse autoencoders are usually used to constrain random features, so that the originally completely random data features become a set of tight and sparse features. Specifically, the sparse feature learning model has become a universal model for researchers to study essential features.[10,16,17]

To extract sparse features from the given training data, formula (15) can be as follows:

$$\arg \min_{\hat{W}} : ||Z\hat{W} - X||_2^2 + \lambda||\hat{W}||_1, \tag{15}$$

where $\hat{W}$ is the solution of the sparse autoencoder, Z is the expected output of the given linear equation, and $XW = Z$.

The above problem is expressed as a convex problem with the lasso in Ref. 18. Therefore, the approximation problem in Eq. (15) can be achieved through orthogonal matching pursuit,[19] K-SVD,[20] Alternating Direction Method of Multipliers (ADMM),[13] fast iterative shrinking threshold Algorithm (FISTA),[14] and other dozens of methods to solve. In addition, existing studies have shown that for the problems involved in the L_1 norm, many of the most advanced algorithms have evolved through ADMM and have also been calculated by it.[15] For example, FISTA is an advanced variant of ADMM. Therefore, the original lasso algorithm is briefly introduced in what follows.

First of all, formula (15) can be equivalent to the general problem shown in formula (16) as follows

$$\arg \min_{W} : f(\omega) + g(\omega), \omega \in R^n, \tag{16}$$

where $f(\omega) = ||Z\omega - x||_2^2$ and $g(\omega) = \lambda||\omega||_1$. According to the ADMM form, formula (16) can be expressed as

$$\arg \min_{W} : f(\omega) + g(\omega), s.t. \omega - o = 0. \tag{17}$$

Equation (17) can be solved by the following iterative steps:

$$\begin{cases} \omega_{k+1} := (Z^T Z + \rho I)^{-1}(Z^T x + \rho(o^k - \mu^k)) \\ o_{k+1} := S_{\frac{\lambda}{\rho}}(\omega_{k+1} + \mu_k) \\ \mu_{k+1} := \mu_k + (\omega - o_{k+1}) \end{cases}, \tag{18}$$

where $\rho > 0$, S is the soft threshold operator, expressed as

$$
\begin{cases}
a - k, & a > k \\
0, & |a| \leq k \\
a + k, & a < -k
\end{cases}
\quad . \tag{19}
$$

2.3. *Singular value decomposition*

Singular value decomposition (SVD) is an important part of linear algebra, that is, any $n \times m$ matrix can be decomposed into

$$
A = U \sum V^T, \tag{20}
$$

where U is an $m \times m$ orthogonal matrix, its column is the eigenvector of $A^T A$, V is an $n \times n$ orthogonal matrix, its column is the eigenvector of $A^T A$, $\sum$ is an $m \times n$ diagonal matrix, It can be expressed as

$$
\sum = diag\{\sigma_1, \ldots, \sigma_r, 0, \ldots, 0\}, \tag{21}
$$

where $\sigma_1 > \sigma_2 > \cdots > \sigma_r$, $r = rank(A)$, and $\sigma_1, \sigma_2, \ldots, \sigma_r$ are the square roots of the eigenvalues of $A^T A$, they are called Singular values of matrix A. Therefore, we have achieved the decomposition of a matrix A, which is one of many effective numerical analysis tools for analyzing matrices. In the broad learning system, two different methods of reducing the size of the matrix are involved. In actual practice, according to different needs, both of these situations may occur. SVD technology is known for its advantages in feature selection.

3. Method

This section introduces the details of methods used in this chapter, including BLS, GAF and the combination system, namely GAF–BLS.

3.1. *Broad learning system*

The structure of BLS is shown in Fig. 3. The input data X of BLS is first converted into mapped features (MF), then MF becomes enhancement nodes (EN) through mapping, and finally, MF and EN

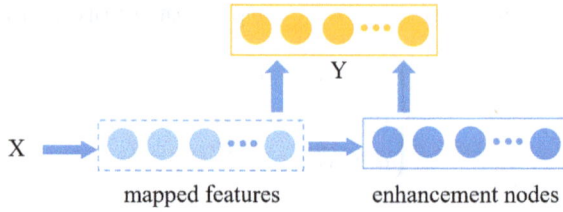

Fig. 3:   Broad learning network.

are connected to the output layer. The difference between BLS and RVFL neural network is that the input data of the RVFL neural network is not mapped into MF but directly connected to the output layer. Therefore, BLS is more flexible in structure than the RVFL neural network, has a stronger ability to handle high-dimensional data than the RVFL neural network, and has a better generalization capabilities. In BLS, the input data first undergo nonlinear mapping of formula (22), and the MF node Z_j can be obtained as follows:

$$Z_i = \phi(XW_{ei} + \beta_{ei}), \quad i = 1, \ldots, n, \tag{22}$$

where Z^n represents all MF nodes, $Z^n = [Z_1, \ldots, Z_n]$. $\phi(\cdot)$ represents a nonlinear function, and W_{ei}, β_{ei} is the randomly generated weight and bias, respectively. Then Z^n is nonlinearly mapped to H_j through formula (23).

$$H_j = \sigma(Z^n W_{hj} + \beta_{hj}), \quad j = 1, \ldots, m, \tag{23}$$

where H_j stands for EN, $\sigma(\cdot)$ stands for nonlinear function, $H^m = [H_1, \ldots, H_m]$ stands for all EN, and W_{hj}, β_{hj} is the randomly generated weight and bias, respectively. Chen *et al.*[10] use linear sparse autoencoders to fine-tune W_{ei} to enhance the feature expression ability of MF. The output layer of BLS is connected to MF and EN at the same time, and the calculation formula is

$$O = [Z^n | H^m]W^o, \tag{24}$$

where O is output response of BLS, and W^o is the weight of the output layer. Furthermore, the objective function of BLS is

$$\min_{W^o} ||O - Y||_2^2 + \lambda ||W^o||_2^2. \tag{25}$$

There are two items in formula (25), the first item is the empirical risk item, Y is the given supervision information, the first item is to reduce the difference between the output of BLS and the supervision information Y, and the second item is the structural risk item which is used to improve the generalization ability of BLS and reduce the risk of over-fitting. λ is the coefficient.

The algorithm of a broad learning system includes two kinds, linear sparse autoencoder and ridge regression theory.

Linear Sparse Autoencoder (LSAE)[10]: To improve the sparse expression ability of the input data, BLS uses LSAE to fine-tune the weight W_{ei} to improve the feature expression ability of the input data. For the input data X, LSAE solves the following optimization problem:

$$\arg\min_{W^*} ||ZW^* - X||_2^2 + \lambda ||W^*||_1, \tag{26}$$

where W^* represents the optimal solution, and Z is the sparse representation of the input data. Equation (26) can be solved by a variety of methods, such as orthogonal matching pursuit,[14,15] ADMM,[10,16] and fast iterative-shrinking threshold algorithm.[17] Equation (26) can also be written in the following form:

$$\arg\min_{W} f(w) + g(w), \tag{27}$$

where $f(w) = ||Zw - x||_2^2$, $g(w) = ||\lambda||w||_1$. Further, according to the ADMM algorithm, formula (27) can be written as

$$\arg\min_{W} f(w) + g(o), s.t. w - o = O, \tag{28}$$

where o is the introduced auxiliary variable. Furthermore, formula (28) can be solved according to formula (29):

$$\begin{cases} w_{k+1} = (Z^T Z + \rho I)^{-1}(Z^T x + \rho(o_k - \mu_k)) \\ o_{k+1} = S_{\frac{\lambda}{\rho}}(w_{k+1} + \mu_k) \\ \mu_{k+1} = \mu_k + (w_{k+1} - o_{k+1}) \end{cases}, \tag{29}$$

where $\rho > 0$, $S_{\frac{\lambda}{\rho}}$ is the soft threshold operation under the parameter $\frac{\lambda}{\rho}$, which is defined as

$$S_{\frac{\lambda}{\rho}}(\alpha) = \begin{cases} \alpha - \lambda, & \alpha > \frac{\lambda}{\rho} \\ 0, & |\alpha| \leq \frac{\lambda}{\rho} \\ \alpha + \frac{\lambda}{\rho}, & \alpha < \frac{\lambda}{\rho}. \end{cases} \tag{30}$$

Ridge regression theory[17]: Ridge regression theory is used to solve the weights of the network output layer. Formula (25) is the objective function of BLS. We can calculate the approximate form of the generalized Moore–Penrose inverse by adding a positive number close to 0 on the diagonal of the matrix $[Z|H]^T[Z|H]$ or $[Z|H][Z|H]^T$, as follows:

$$[Z|H]^\dagger = \lim_{\lambda \to 0}(\lambda I + [Z|H][Z|H]^T)[Z|H]^T. \tag{31}$$

Then, the solution of the BLS objective function is

$$W^o = (\lambda i + [Z|H][Z|H]^T)[Z|H]^T Y. \tag{32}$$

3.2. *Gramian angular field*

The GAF[1] evolved on the basis of the Gram matrix. It uses the polar coordinate system to replace the typical Cartesian coordinate system to represent the time series. In GAF, each element is the cosine value of the sum of the paired time series values in the polar coordinate system. As GAF retains the dependence and correlation of the time series well and conforms to the characteristics of time series, GAF it is considered as a way to preserve the correlation between the amplitudes of each time point in the time series. The conversion steps from the original time series to GAF are as follows.

For a time series with q channels and n observations per channel,

$$S = \{S_1, S_2, \ldots, S_q\}, \quad S_i = \{s_{i1}, s_{i2}, \ldots, S_{in}\}. \tag{33}$$

First, normalize it by channel:

$$\widetilde{s}_{ij} = \frac{(s_{ij} - \max(S_i)) + (s_{ij} - \min(S_i))}{\max(S_i) - \min(S_i)}. \tag{34}$$

Make all values fall within $[-1, 1]$. Then, using polar coordinates to retain the absolute time relationship of the sequence, the value is encoded as an angular cosine, and the timestamp is encoded as a radius, as follows:

$$\phi = \arccos(\widetilde{s}_{ij}), -1 < \widetilde{s}_{ij} < 1, \widetilde{s}_{ij} \in \widetilde{S}_i, \widetilde{S}_i \in \widetilde{S}, r_{ij} = \frac{t_j}{N}, t_j \in N \tag{35}$$

where $\widetilde{s}_{ij}$ is an observation in $\widetilde{S}_i$, t_j is the timestamp corresponding to the time series $\widetilde{S}$, and N is the total length of the timestamp.

The time series $\widetilde{S}$ is encoded in polar coordinates, and is bijective, the cosine value of interval $[-1, 1]$ is monotonous in the polar coordinate angle range $[0, \pi]$, and has a unique inverse. As time increases, the lines of corresponding values bend between different angle points on the polar coordinate circle. After converting the rescaled time series to the polar coordinate system, the following operation is defined:

$$x \bigoplus y = \cos(\theta_1 + \theta_2), \qquad (36)$$

where θ_1 and θ_2 represent the corresponding angles of the vectors in the polar coordinate system, respectively, and the definition of the GAF of the one-dimensional time series is as follows:

$$\widetilde{G} = \begin{pmatrix} \cos(\phi_1 + \phi_1) & \cos(\phi_1 + \phi_2) & \cdots & \cos(\phi_1 + \phi_n) \\ \cos(\phi_2 + \phi_1) & \cos(\phi_2 + \phi_2) & \cdots & \cos(\phi_2 + \phi_n) \\ \vdots & \vdots & & \vdots \\ \cos(\phi_n + \phi_1) & \cos(\phi_n + \phi_2) & \cdots & \cos(\phi_n + \phi_n) \end{pmatrix}. \qquad (37)$$

In GAF, the diagonal is composed of the original values of the time series after scaling, and the time increases with the diagonal from the upper left corner to the lower right corner, so the time dimension is correspondingly encoded into the geometric structure of the matrix. $\widetilde{G}_{(i,j\|i-j|=k)}$ expresses the relative correlation through superposition in the direction of the corresponding time interval k. The main diagonal $G_{(i,j)}$ is a special case when $k = 0$.

3.3.   GAF–BLS

As shown in Fig. 4, based on the above-mentioned BLS and GAF algorithms, this chapter proposes a time series classification method based on GAF and BLS as follows: (1) Convert each sample into a feature matrix according to the GAF encoding method; (2) Input the feature matrix into the built broad learning system and tune the parameters so that it can adaptively extract the relevant information in the feature matrix, learn the information of each category, and update the network weights; (3) Use the trained broad learning network to classify time series data.

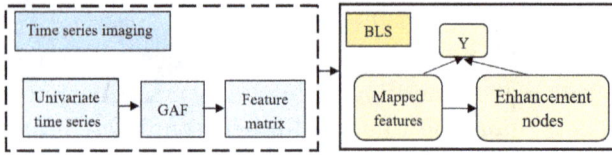

Fig. 4:   The workflow of GAF–BLS.

This method has the following advantages:

(1) Feature matrix generated by the GAF method replaces the univariate time series, which can better explain and present the commonality and potential relationships of time series data.
(2) BlS has done further processing on the feature matrix, which not only greatly increases the feature extraction speed, but also improves the classification accuracy.
(3) The pseudo-inverse method is used to solve the weights, which avoids the gradient explosion problem and iterative process in the gradient descent method, and reduces the time cost.

3.3.1.   Feature matrix generation

Since univariate time series have only a one-way correlation in the horizontal and vertical directions, the potential relationship of the data cannot be explained to a certain extent, and the dot product operation of one-dimensional series cannot distinguish between the effective information in the signal and Gaussian noise, so we map univariate time series data to polar coordinates. As the sequence increases with time, the corresponding values will be distorted between different points on the polar coordinates, just like water ripples. The sequence generated in this way is bijective, no information is lost, and time dependence is maintained. This representation based on polar coordinates is a new way to understand one-dimensional signals. Afterward, GAF is generated using angular perspective. By changing the spatial dimension, the angular relationship can be fully utilized to dig out hidden features in the disturbance signal, and the processed sparse output can easily distinguish effective information from Gaussian noise. BLS is susceptible to noise and its ability to deal with sparse data has been fully confirmed. Therefore, mapping univariate time series to polar coordinates and generating GAF, and then using BLS to learn it, can effectively improve the accuracy of classification.

3.3.2. *Broad learning network training*

The training process is described according to the broad learning network structure of Fig. 3. Firstly, the original input layer obtains the mapping features of the data through the sparse coding algorithm. Secondly, the mapping features as a feature node layer are connected to the enhancement layer through an activation function, and the weights in it are randomly generated. Finally, all mapping features and enhancement nodes are directly connected to the output as input. To find the required connection weight, ridge regression[21] improved algorithm of the least square method[22] is used to obtain the weight between the input layer and the output layer.

4. Experiment and Discussion

To test the effectiveness of the GAF–BLS method on the time series classification problem, we use three datasets from the UCR time series archive to compare the classification ability and efficiency of GAF–BLS and several commonly used time series classification algorithms. The three datasets are Beef, ClorineConcentration, and ECG5000. The details of the datasets are shown in Table 1.

The recognition rate and recognition speed are adjusted through the relevant parameter settings. The main parameters include the following aspects: the number of feature mapping nodes in each group

Table 1: Information of datasets.

Datasets	Type	Length	No. of classes	Training size	Test size
Beef	Spectro	470	5	30	30
Clorine Concentration	Simulated	166	3	467	3,840
ECG5000	ECG	140	5	500	4,500

Table 2:   Experimental analysis of GLS-BLS based on Beef, ClorineConcentration, and ECG5000.

Datasets	N_f	N_m	N_e	Training time/s	Accuracy/%
Beef	90	1	100	1.3645	93.33
ClorineConcentration	118	13	1	2.0740	95.55
ECG5000	9	3	40	0.0957	94.47

N_f, the number of feature mapping groups N_m, and the number of enhancement nodes N_e. Adjust the above parameters through grid search to obtain the best results and achieve high-accuracy recognition results. Table 2 shows the parameter settings and experimental results of GAF–BLS to achieve the highest accuracy on the three datasets of Beef, ClorineConcentration, and ECG5000.

It can be seen from Table 2 that the accuracy of GAF on these time series datasets is more than 90%, and the training time is within 3 s. This not only proves the effectiveness of the GAF–BLS method but also conforms to our opinion on GAF–BLS expectation of a high recognition rate and short training time.

In addition, we also compare GAF–BLS method with several commonly used time series methods, and calculate the classification accuracy and training time of different classification algorithms on these datasets. And in each dataset, we compare the accuracy of all algorithms with the training time, and highlight the best results in bold. Table 3 shows the performance of different algorithms on different datasets. In terms of accuracy, GAF–BLS is significantly higher than other algorithms on the Beef and ClorineConcentration datasets, and the accuracy on the ECG5000 dataset is lower than the ROCKET algorithm by 0.27%. In terms of training time, the training efficiency of GAF–BLS is higher than most deep learning algorithms, and only ROCKET is basically the same. From the perspective of datasets, GAF–BLS has higher accuracy on Beef and ClorineConcentration datasets with smaller sample sizes and is more suitable for classification problems of small samples. In general, the

Table 3: The performance of different algorithms on Beef, ClorineConcentration, and ECG5000.

	Accuracy/%			Training time/s		
Algorithms	Beef	Clorine	ECG	Beef	Clorine	ECG
InceptionTime	73.33	87.70	94.35	392.54	892.88	1,005.99
MCNN	20	53.25	58.37	67.84	229.82	255.53
tlenet	70	79.08	93.53	1,169.03	752.35	789.31
twiesn	66.66	55.62	92.2	0.81	4.32	3.18
mcdcnn	66.66	64.63	**94.11**	6.74	16.85	16.67
cnn	80	59.81	92.73	122.06	1,048.35	1,364.81
ROCKET	83.66	81.21	94.74	**0.33**	**1.37**	1.25
FCN	73.33	82.31	94.2	222.47	1,505.65	1,912.77
MLP	70	55.96	94.04	383.64	2,464.53	3,294.55
RESNET	80	85.20	93.73	294.79	888.092	715.39
GAF–BLS	**93.33**	**95.55**	94.47	1.36	2.0740	**0.09**

Note: The bold entries represent the best values.

GAF–BLS algorithm performs better than most deep learning algorithms. Although compared with the ROCKET algorithm it has not achieved complete victory, the network structure of GAF–BLS is simpler and more interpretable.

5. Conclusion

This chapter proposes the GAF–BLS algorithm based on the GAF and BLS to solve the time series classification problem. The algorithm uses the GAF to convert the time series data into feature matrices. The GAF performs well in extracting the time series characteristics of the time series data, and can retain the time series characteristics to the greatest extent. Then, it uses the feature matrices mapped by broad learning for model training. The network structure of this model makes full use of the ability of BLS to quickly process complex data. Its advantages are simple structure and short model training time. Experimental results prove that GAF–BLS has a simpler model structure and smaller parameter scale, and can outperform most deep learning models in terms of training speed and classification accuracy on time series datasets.

References

1. Z. Wang and T. Oates, Imaging time-series to improve classification and imputation, in *Twenty-Fourth International Joint Conference on Artificial Intelligence*, 2015, pp. 3939–3945.
2. N. Hatami, Y. Gavet, and J. Debayle, Classification of time-series images using deep convolutional neural networks, in *Tenth International Conference on Machine Vision (ICMV 2017)*, 2018, vol. 10696, p. 106960Y.
3. W. Hou, X. Gao, D. Tao, and X. Li. Blind image quality assessment via deep learning, *IEEE Transactions on Neural Networks and Learning Systems*. **26**(6), 1275–1286 (2014).
4. J. Li, J.-H. Cheng, J.-Y. Shi, and F. Huang, Brief introduction of back propagation (BP) neural network algorithm and its improvement, in *Advances in Computer Science and Information Engineering*, 2012, pp. 553–558, Springer.
5. L. Deng and D. Yu, Deep learning: Methods and applications, *Foundations and Trends in Signal Processing*. **7**(3–4), 197–387 (2014).
6. Y. LeCun, L. Bottou, Y. Bengio, and P. Haffner, Gradient-based learning applied to document recognition, *Proceedings of the IEEE*. **86**(11), 2278–2324 (1998).
7. K. Simonyan and A. Zisserman, Very deep convolutional networks for large-scale image recognition, arXiv preprint arXiv:1409.1556 (2014).
8. G. E. Hinton, Deep belief networks, *Scholarpedia*. **4**(5), 5947 (2009).
9. N. Srivastava and R. R. Salakhutdinov, Multimodal learning with deep Boltzmann machines, *Advances in Neural Information Processing Systems*. **25**, 2222–2230 (2012).
10. C. P. Chen and Z. Liu, Broad learning system: An effective and efficient incremental learning system without the need for deep architecture, *IEEE Transactions on Neural Networks and Learning Systems*. **29**(1), 10–24 (2017).
11. C. P. Chen and Z. Liu, Broad learning system: A new learning paradigm and system without going deep, in *2017 32nd Youth Academic Annual Conference of Chinese Association of Automation (YAC)*, 2017, pp. 1271–1276.
12. C. P. Chen, Z. Liu, and S. Feng, Universal approximation capability of broad learning system and its structural variations, *IEEE Transactions on Neural Networks and Learning Systems*. **30**(4), 1191–1204 (2018).
13. S. Boyd, N. Parikh, E. Chu, B. Peleato, J. Eckstein, *et al.*, Distributed optimization and statistical learning via the alternating direction method of multipliers, *Foundations and Trends® in Machine Learning*. **3**(1), 1–122 (2011).

14. A. Beck and M. Teboulle, A fast iterative shrinkage-thresholding algorithm for linear inverse problems, *SIAM Journal on Imaging Sciences.* **2**(1), 183–202 (2009).

15. T. Goldstein, B. O'Donoghue, S. Setzer, and R. Baraniuk, Fast alternating direction optimization methods, *SIAM Journal on Imaging Sciences.* **7**(3), 1588–1623 (2014).

16. Z. Liu, J. Zhou, and C. P. Chen, Broad learning system: Feature extraction based on K-means clustering algorithm, in *2017 4th International Conference on Information, Cybernetics and Computational Social Systems (ICCSS)*, 2017, pp. 683–687.

17. M. Xu, M. Han, C. P. Chen, and T. Qiu, Recurrent broad learning systems for time series prediction, *IEEE Transactions on Cybernetics.* **50**(4), 1405–1417 (2018).

18. R. Tibshirani, Regression shrinkage and selection via the lasso, *Journal of the Royal Statistical Society: Series B (Methodological).* **58**(1), 267–288 (1996).

19. J. A. Tropp and A. C. Gilbert, Signal recovery from random measurements via orthogonal matching pursuit, *IEEE Transactions on Information Theory.* **53**(12), 4655–4666 (2007).

20. M. Aharon, M. Elad, and A. Bruckstein, K-SVD: An algorithm for designing overcomplete dictionaries for sparse representation, *IEEE Transactions on Signal Processing.* **54**(11), 4311–4322 (2006).

21. A. E. Hoerl and R. W. Kennard, Ridge regression: Biased estimation for nonorthogonal problems, *Technometrics.* **12**(1), 55–67 (1970).

22. Y. LeCun, F. J. Huang, and L. Bottou, Learning methods for generic object recognition with invariance to pose and lighting, in *Proceedings of the 2004 IEEE Computer Society Conference on Computer Vision and Pattern Recognition, 2004. CVPR 2004*, 2004, vol. 2, pp. II–104.

Chapter 7

Denoising of Radio Modulation Signal Based on Deep Learning

Hongjiao Yao, Qing Zhou, Zhuangzhi Chen, Liang Huang, Dongwei Xu, and Qi Xuan*

*Institute of Cyberspace Security,
Zhejiang University of Technology,
Hangzhou, P.R. China*

**xuanqi@zjut.edu.cn*

In the actual environment, the communication process will be affected by noise inevitably. How to reduce the influence of noise is of great significance to improve communication quality. Traditional signal denoising techniques work with weak generalization and need some prior information of noise to map the received signal to a separable transform domain for separation. Therefore, in this chapter, we propose a radio modulation signal denoising technology based on deep learning, that is, an end-to-end generative model, which makes full use of deep learning to automatically extract features and simplify the signal denoising step. The generative model is based on a generative adversarial network, and its loss function adds average error loss and continuity loss in line with signal characteristics based on traditional LSGAN loss. Experiments have proved that this method is better than the low-pass filter in the time and frequency domain, has good generalization ability, and can denoise different modulation signals.

1. Introduction

Signals are widely used in the field of communication, but radio communication channels are usually complex and will cause certain interference to communication. Therefore, signal denoising has always been a hot research topic. The existing denoising methods are mainly divided into two categories: traditional methods and learning-based methods. The traditional methods in the past mainly included nonlearning methods such as filtering, wavelet transform, and empirical mode decomposition. Linear denoising methods that rely on filtering, such as Wiener filters,[1] work well in the presence of stationary noise, but they show limitations when the signal and noise share the same frequency spectrum.[2] In wavelet transform, the signal is divided into different proportions according to the frequency range, and the noise is filtered through a reasonable threshold,[3-6] which lacks the ability of an adaptive decomposition layer.[7] Finally, empirical mode decomposition[8] is a data-driven method, suitable for stationary and non-stationary signals, but it is difficult to decompose the signal into unique frequency components, resulting in mode mixing.[9]

Learning-based denoising methods are particularly popular in the image field, mainly including encoder and decoder networks,[10] deep denoising networks based on residual learning,[11] and multi-level wavelet neural networks[12] methods. Similarly, in audio and speech processing, deep neural networks[13,17] have also made some progress. In addition, there are also some for seismic noise signals,[18,19] electrocardiograms and motion signals,[20,22] gravitational waves. Signal[23] is based on the study of learning methods, but the research on radio modulation signals is quite limited.[24]

Based on the study of deep learning methods for image denoising and speech denoising, we propose a deep denoising network structure based on a generative adversarial network for adaptive denoising. The main contributions of our work are as follows:

(1) We propose a new signal denoising method based on generative adversarial network. This method does not require prior knowledge of signals, and adaptive denoising can be achieved through learning.
(2) The proposed denoising method can learn the signal characteristics of nine modulation-types at the same time, and realize the

denoising function for each modulation-type signal separately. Within a certain range, input noise signals with different signal-to-noise ratios can also have denoising effects.

(3) We found that by learning from adversarial samples and original samples, the denoising network can also achieve a certain defensive effect.

(4) When analyzing the denoising effect, we combined constellation diagram, visualization, and time–frequency domain analysis for the first time.

The rest of this chapter is organized as follows. In Section 2, we introduced the model and receiving system for generating radio modulation signals. In Section 3, we introduce the technical route and details of our proposed method. In Section 4, we describe the experiment in detail, discuss and analyze the results. We finally draw a brief conclusion in Section 5.

2. Preliminary

2.1. *GAN*

GAN contains two network models, one is a generative model and the other is a discriminative model. The two are trained alternately to fight against learning. The generative model is learning how to map the distribution Z from a certain sample z to the distribution X from another sample x, where sample x is the training sample and sample z is the generated pseudo sample. The neural network that performs the mapping in the generative adversarial network structure is called the generator (G), and its main task is to learn an effective mapping that can imitate the real data distribution to generate new samples related to the samples in the training set. The important thing is that the function of the generator is not realized by memorizing the input and output pairs, but by learning the data distribution characteristics and then mapping to the sample z.

The generator learns to learn the mapping through adversarial training, and there is another neural network called the discriminator (D). The discriminator is usually a classifier, and its input has two cases, one is a real sample, from the dataset that the generator is imitating, and the other is a pseudo sample generated by the generator.

The adversarial feature is embodied in that the discriminator must classify the samples from the dataset as real samples, and the sample $G(z)$ from the generator must be classified as false. The generator tries to deceive the discriminator by adjusting its parameters, allowing the discriminator to classify the generator's output as real. In the backpropagation process, the discriminator tries to perform better on the real features in the input. At the same time, the generator corrects its parameters to approach the real data distribution obtained from the training data and adjusts against the other. This adversarial learning process can be regarded as a minimax game between the generator and the discriminator. The objective function is shown in formula (1), where $D(x)$ represents the distribution of real data, and $G(z)$ represents the distribution of generated data.

$$\min_{G} \max_{D} V(D, G) = E_{x \sim P_{\text{data}}(x)}[\log D(x)]$$
$$+ E_{z \sim P_z(z)}[\log(1 - D(G(z)))]. \tag{1}$$

2.2. CGAN

CGAN is an extension of GAN. The main difference between the two is that both the generator and the discriminator add additional information y as a condition, and y can be any information, such as category information and data information. The reason for adding conditional information is that GAN learns directly on the data distribution, and the optimization direction is difficult to control. CGAN inputs the prior noise z and condition information y into the generator and discriminator for training. In the generator, *a priori* noise z and conditional information y are connected to form joint implicit characterization information, which guides the signal generated by the generator.

Generating data and generating labels need to pass the judgment of D at the same time, so the loss function of CGAN can be expressed as

$$\min_{G} \max_{D} V(D, G) = E_{x \sim P_{\text{data}}(x)}[\log D(x \mid y)]$$
$$+ E_{z \sim P_z(z)}[\log(1 - D(G(z \mid y)))] \tag{2}$$

2.3. *DCGAN*

DCGAN mainly optimizes the GAN network structure and uses convolution in GAN. Before the specific introduction, first, explain that the function of convolution is to extract some specific features in the data, different numbers of convolution kernels, the extracted features of which are different; the extracted features are the same, and the different convolution kernels have the same effect. Not the same. The main improvement of DCGAN is that the spatial pooling layer is replaced with a convolutional layer with specified step size, the generator and discriminator both use batch normalization (BN), the fully connected layer is removed, and the generator uses ReLU as the activation function (Except the output layer), all layers of the discriminator use Leaky ReLU as the activation function.

The reasons for the improvement are explained as follows: First of all, BN has data with a unified specification, which makes it easier for the model to learn the laws between the data, thereby helping the network to converge. Furthermore, the use of transposed convolution allows the neural network to learn to upsample in the best way, and the use of convolution makes downsampling no longer a fixed discard of pixel values in certain locations but lets the network learn to downsample by itself this way. At the same time, removing the fully connected layer is global mean pooling, although it helps the stability of the model it reduces the convergence speed. Finally, the Leaky ReLu activation function has a fast calculation speed and solves the problem of GAN's easy gradient disappearance.

3. Methods

3.1. *SDGAN*

The specific experimental flow of the radio modulation signal deep denoising method based on supervised learning is shown in Fig. 1.

The radio modulation signal deep denoising model based on supervised learning is a generative countermeasure network used for signal denoising, which combines the characteristics of the above-mentioned different GANs for fusion. The deep denoising model we

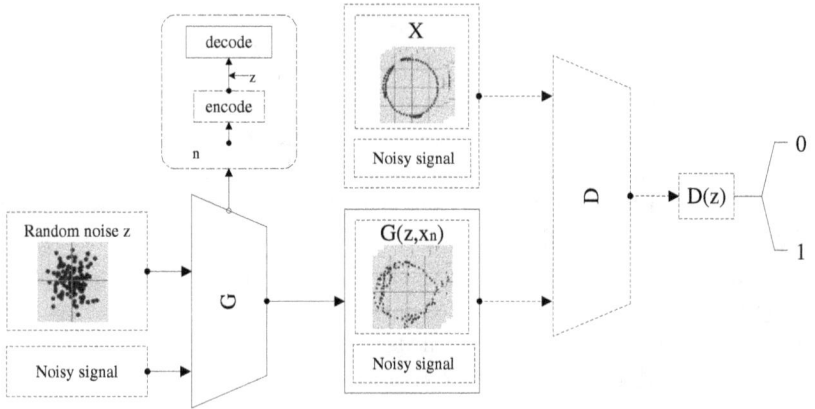

Fig. 1:   The learning process of the SDGAN model.

designed also has a generator and a discriminator. Moreover, the generator and discriminator are designed based on DCGAN. The generator structure is first convolved and then deconvolved. The data input to the generator is noisy data and random noise z. The input random noise z is to allow the generator to better learn the pure signal data distribution, and it is added in the high-dimensional space after convolution. Compared to adding it in the low-dimensional space, the model convergence rate can be improved. The input of the discriminator is designed based on CGAN.

By adding the noisy signal as a label, the data distribution of the noisy signal can match the corresponding modulation type while learning how to approach the data distribution of the pure signal. The resulting objective function is as shown in formula (3), where the input representing the discriminator is a data pair of pure signal and noise-containing signal, and $D\left(G\left(z, x_n\right), x_n\right)$ representing the discriminator's input is the denoising signal and noise-containing signal generated by the generator data pairs.

$$\min_{G} \max_{D} V(D, G) = E_{x, x_n \sim P_{\text{data}}\left(x, x_n\right)} \left[\log D\left(x, x_n\right)\right]$$
$$+ E_{z \sim P_z(z) x_n - p_{\text{data}}(x_n)} \tag{3}$$
$$\times \left[\log\left(1 - D\left(G\left(z, x_n\right), x_n\right)\right)\right].$$

3.2. *Generator and a discriminator design*

The training phase of the signal denoising network structure we designed requires a generator and a discriminator. The input of the generator is a noisy signal and random noise z. We designed the generator into a fully convolutional structure so that the network can concentrate on learning the input data and the context information of each level in the entire convolution and deconvolution process. We have reduced the time required for training. The convolution kernel of all convolutions is 3, the step size is 2, and the padding is 1; the convolution kernel of all inverse convolutions is 4, the step size is 2, and the padding is 1. In the convolution stage, the input signal is continuously compressed through six one-dimensional convolutional layers with parameter correction linear units (PReLUs) to obtain a compressed representation vector. In the decoding stage, the compressed representation vector and the random noise of the same dimension represented by N in Fig. 6 are connected and input to the inverse convolution layer, and the original data length is restored through a total of six one-dimensional inverse convolutions. The generator also has a jump connection structure, connecting each convolutional layer to its homologous deconvolutional layer, bypassing the compression performed in the middle of the model. Doing so allows the input and output of the model to share the same underlying structure so that the low-level details of the signal can be passed to the deconvolution stage.

As shown in Fig. 2, which is the structure diagram of the generator model we designed, after the training is completed, the generator extracted can be used as a denoising model, with the input being the noisy signal, and the output is the desired denoising signal.

Since the generator and the discriminator in the generative adversarial network are against each other, we design the discriminator structure to be similar to the convolution part of the generator, as shown in Fig. 3. The difference lies in two points. The first is that in the discriminator, we use Virtual Batch Norm to sample the reference batch before training. Usually, in the forward pass, we can pre-select the reference batch to calculate the normalization parameters (μ and σ) of the batch normalization (BN). However, since we

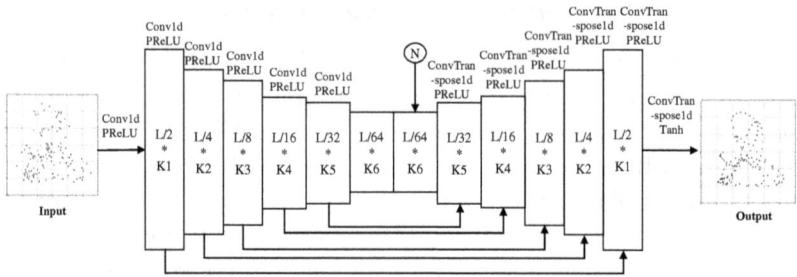

Fig. 2:   The generator structure diagram in the deep denoising model.

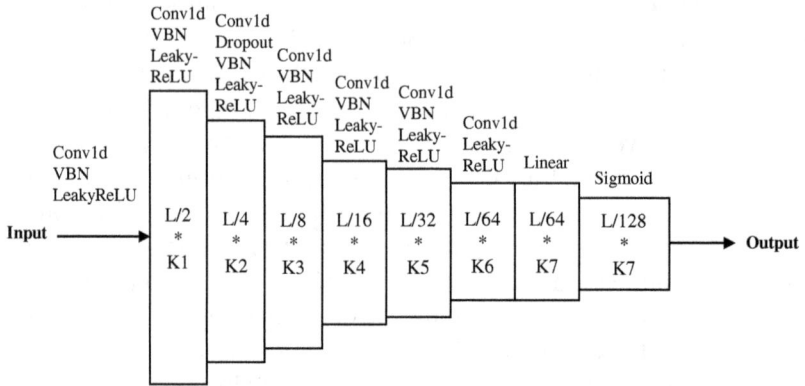

Fig. 3:   The structure diagram of the discriminator in the depth denoising model.

use the same batch throughout the training, it is possible to overfit this reference batch. To alleviate this situation, VBN can combine the reference batch with the current batch to calculate the normalization parameters. The second point is that the dropout operation is added after the third layer of convolution to avoid overfitting.

3.3.   Loss function design

One of the characteristics of the generator is its end-to-end structure, which does not require any intermediate transformation for extracting signal features, and the optimization direction is determined by the loss function of the discriminator and the generator. However, in the course of the experiment, we found that it is easy to encounter the problem of gradient disappearance when using cross-entropy loss in training.

Therefore, to solve this problem, we consider changing the loss function to a minimum mean square error. The minimum mean square error will gradually decrease as the loss function converges, but this will also cause a greater impact on model parameter adjustment when the model encounters an abnormal point, resulting in a decrease in model training efficiency. So we also considered the average absolute error, which is milder than the mean square error when dealing with abnormal points, but its derivative is V-shaped, which is not conducive to convergence. Finally, we found that there are precedents in the field of GAN research that use Least Squares Loss (LSGAN) instead of cross-entropy loss to solve the two problems of low quality and unstable training. The objective function is designed as follows:

$$\min_{D} V_{\mathrm{LSGAN}}(D) = \frac{1}{2} E_{x,x_n \sim P_{\mathrm{data}}(x,x_n)} \left[(D(x, x_n) - 1)^2 \right]$$
$$+ \frac{1}{2} E_{z \sim P_z(z), x_n \sim p_{\mathrm{data}}(x_n)} \left[D(G(z_x x_n) x_s x_n)^2 \right],$$

$$(4)$$

$$\min_{G} V_{\mathrm{LSGAN}}(G) = \frac{1}{2} E_{z \sim P_z(z), x_n \sim P_{\mathrm{data}}(x_n)} \left[(D(G(z, x_n), x_n) - 1)^2 \right].$$

$$(5)$$

We redesigned the loss function based on LSGAN and combined the characteristics of signal data. Considering that the signal data is continuous in time, we added the continuity loss of the data generated by the generator to the generator loss to constrain. The difference between every two adjacent data it in the generated data such that it is the smallest. At the same time, we found in the training process that it is difficult to train a more accurate model if only the discriminator loss is used in the generator loss, so we added the average absolute error to the generator loss. The final generator loss function is shown in formula (6), and the best depth denoising model is obtained by adjusting the parameters λ_1, λ_2, and p.

$$\min_{G} V_{\mathrm{LSGAN}}(G) = \frac{1}{2} E_{z \sim P_z(z), x_n \sim p_{\mathrm{data}}(x_n)} \left[(D(G(z, x_n), x_n) - 1)^2 \right]$$
$$+ \lambda_1 \| G(z, x_n) - x \|_1 + \lambda_2 \| \nabla \omega \|_p^p.$$

$$(6)$$

4. Experiment

4.1. *Dataset and setting*

The radio modulation signal dataset used in our chapter is self-made based on the public dataset RadioML2016.10a. The following will introduce in detail the three aspects of the public dataset, adding channel noise, and enhancing the dataset. In the end, we got 90,000 samples with Gaussian noise, of which 89,100 are randomly selected as the training set, and the remaining 900 are used as the test set. In addition, we designed an interference dataset with a total of 80,000 samples, of which 79,200 were randomly selected as the training set, and the remaining 800 were used as the test set.

Through literature research, we have learned that the noise in the channel can be mainly divided into additive noise, narrow-band Gaussian noise, and multiplicative noise. The specific difference is that additive noise is independent of unwanted electrical signals that always exist in the signal. Nature can be further subdivided into impulsive noise (a large amplitude and short duration), narrow-band noise (a sine wave of a single frequency with constant amplitude), and fluctuating noise (random noise generally present in the time-frequency domain); Narrow-band Gaussian noise is generated after fluctuating noise is filtered by a band-pass filter at the receiving end; multiplicative noise does not exist independently, but appears along with signal transmission.

The "noise source" involved in the communication system model refers to the additive noise (mainly fluctuating noise) scattered throughout the communication system, which includes all thermal noise, shot noise, and cosmic noise in the channel. In addition, according to a large number of practices, fluctuating noise is not only a Gaussian noise, but the frequency spectrum is also evenly distributed in a relatively wide frequency range, so fluctuating noise is often called white noise. Therefore, the noise in the communication system is often approximated as Gaussian white noise.

Based on the above information, because the data of different signal-to-noise ratios in the public dataset are not one-to-one, we add quantitative Gaussian white noise to the pure signal as the noise dataset required by the research. Figure 4 shows a constellation diagram for comparison between the transmitted signal and noisy information.

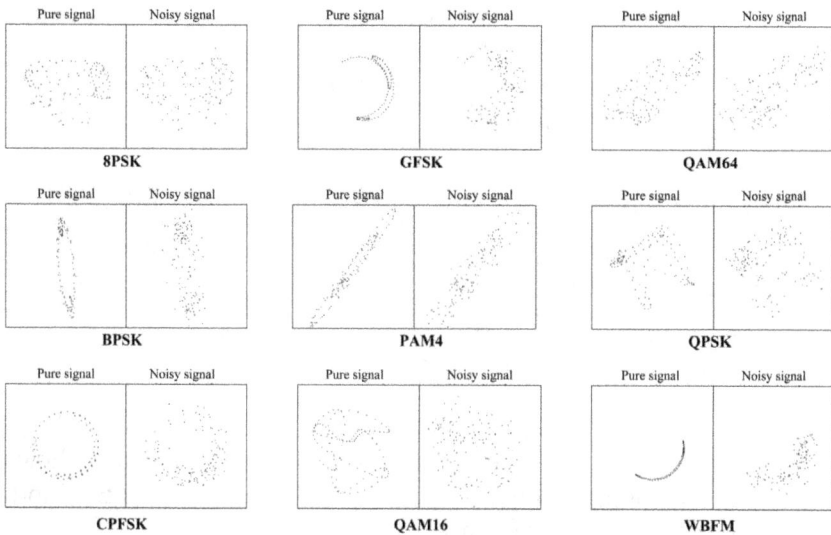

Fig. 4: Constellation diagram for comparison between the transmitted signal and noisy information.

In addition to Gaussian noise in the channel, there may also be interference from other modulated signals in the same channel when the signal is being transmitted. At this time, the transmitted signal is a QPSK modulated signal, the interference source is a BPSK modulated signal, and the final received signal is called the received signal. As a result, we have also produced an interference dataset. The specific steps are as follows:

(1) Select the BPSK modulation type data as the interference source from the pure signal, and the QPSK modulation type data as the transmission signal, each with 1,000 samples.
(2) Multiply the BPSK modulation type data by the coefficients 0.2 and 0.3, and then superimpose them on the QPSK modulation type data to obtain 2,000 samples.
(3) In the actual transmission process, the two types of modulation data may not be transmitted at the same time, so we consider superimposing according to different positions. Since the data length in the public dataset is 128, we finally superimpose the BPSK data from 0, 16, 32, and 64 bits on the QPSK modulation type data to obtain 8,000 samples.

Transmitted signal   Interference signal   Received signal

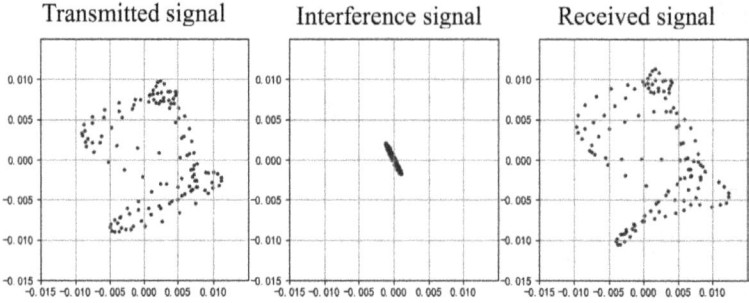

Fig. 5:   Constellation diagram of the comparison of the received signal, the interference signal, and the transmitted signal.

The final comparison constellation diagram of the received signal, the interference signal, and the transmitted signal is shown in Fig. 5. The transmitted signal is a pure signal of QPSK modulation type, the interference signal is a BPSK modulated signal with a coefficient of 0.2, and the received signal is a superimposition of pure signal and the interference signal.

4.2. Evaluation metrics

The constellation diagram is in a coordinate system, the abscissa is I, the ordinate is Q, the projection on the I axis is the in-phase component, and the projection on the Q axis is the quadrature component. Because of the difference in signal amplitude, the signal may fall within the unit circle, not only on the unit circle. Specifically, 16 pulse-amplitude modulation (16PAM), so each symbol needs four binary representations. According to the difference of amplitude and phase, the place where the 16 symbols fall is different. Jumping from one point to another means that the phase modulation and amplitude modulation are completed at the same time. The constellation diagram can completely and clearly show the mapping relationship of digital modulation.

Drawing a constellation diagram can compare the purity of different signals on a two-dimensional plane. The purer the denoising signal obtained by the denoising technology, the lower the bit error rate during decoding, and the higher the communication quality.

Therefore, the use of constellation diagrams can subjectively judge the quality of the denoising method.

This chapter also uses the traditional frequency-domain map when analyzing the denoising effect. The frequency-domain diagram can be used to compare the amplitude of the effective signal and the recovery of the frequency band in which different denoising methods are located.

In addition to comparing the denoising effect in the subjective time–frequency domain, this experiment also proposes a new purity index to objectively compare the purity of the noisy signal, denoised signal, and filtered signal. The higher the value of the purity index, the purer the signal. In the calculation process, the amplitude of the transmitted signal is used as the amplitude of the effective signal to calculate the effective signal power. When calculating the purity index of the noisy signal, the denoised signal, and the filtered signal, the amplitude difference between the noisy signal and the transmitted signal, the amplitude difference between the denoising signal and the transmitted signal, and the amplitude difference between the filtered signal and the transmitted signal are used to calculate the purity index, so as to get the purity of three different signals.

4.3. *Experimental results and discussion*

We use the trained generator model as the denoising model, input the noise signal into the denoising model, and the output is the denoising signal. We analyze the denoising effect from the constellation diagram and the time–frequency domain, respectively. Whether the noise is Gaussian noise or superimposed noise, our denoising method is better than the low-pass filter.

4.3.1. *Experiment on Gaussian noise*

According to Fig. 6, the constellation diagram, it can be seen that the denoising method proposed in this chapter is better than the traditional filtering denoising method, chapter which mainly reflects in (1) Better continuity and better curvature; (2) the degree of gathering is better, and there is no obvious discrete point.

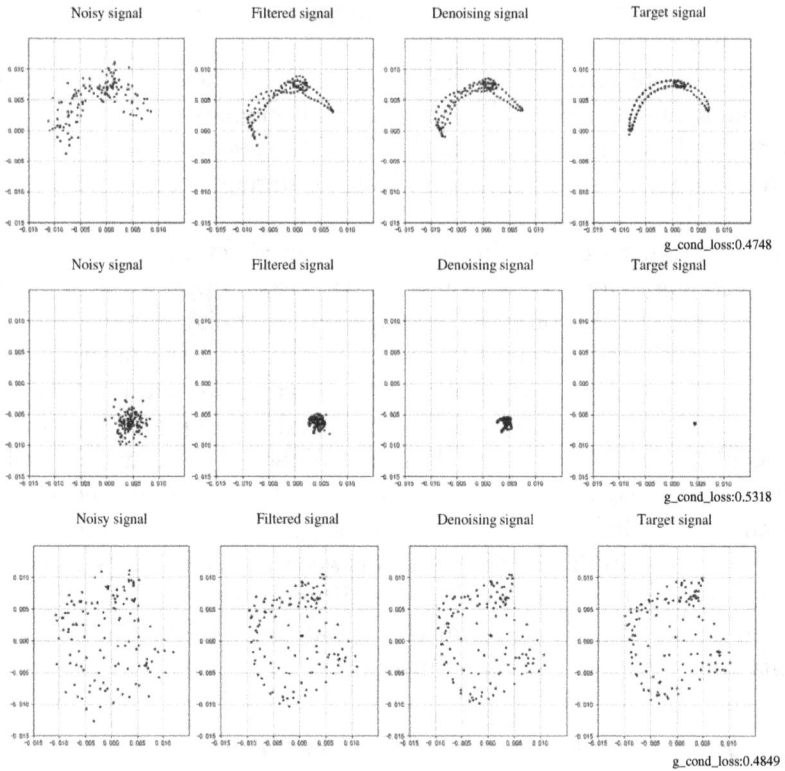

Fig. 6:   Comparison of constellation diagrams for Gaussian noise.

From Fig. 7, the time–frequency domain diagram, it can be seen that the denoising effect of our denoising method in the frequency domain is better than that of the low-pass filter. Especially in the low-frequency part, the signal trend is closer to the pure signal, and more noise is removed.

Figure 8 shows a bar graph comparing the purity indicators of three different signals. From the figure, it can be seen that the purity indicators of all modulation types after denoising are improved by an average of 0.95 dB compared to the purity indicators of the filtered signal. The WBFM modulation type is improved by about 2 dB. This shows that the denoising method mentioned in this chapter also has a better performance in terms of the improvement effect of the purity index.

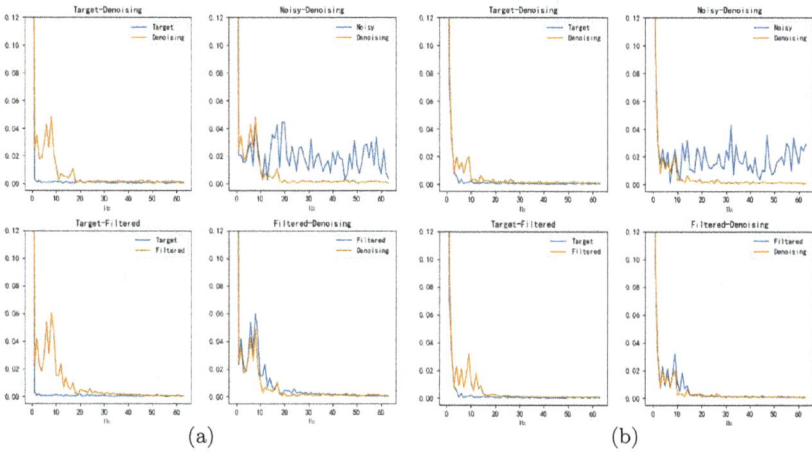

Fig. 7: Time–frequency domain comparison for Gaussian noise.

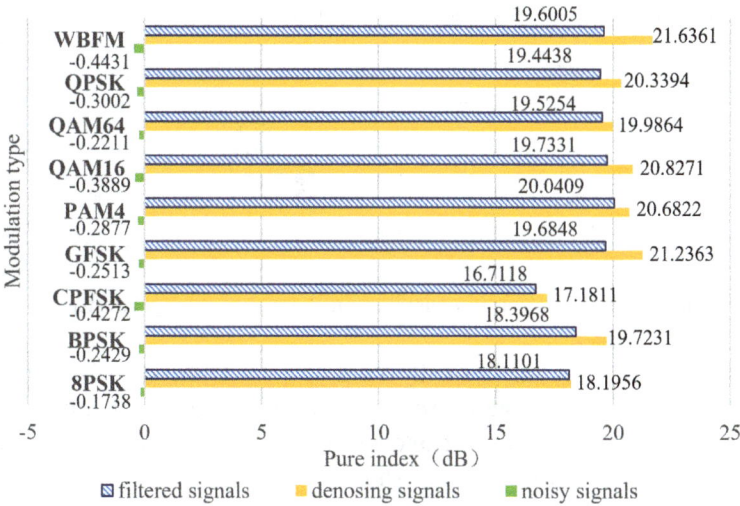

Fig. 8: Pure index comparison bar chart.

4.3.2. *Experiment on superimposed signal*

According to Fig. 9, the constellation diagram, it can be seen that the denoising method proposed in this chapter is better than the traditional filtering denoising method, mainly reflected in the fact that.

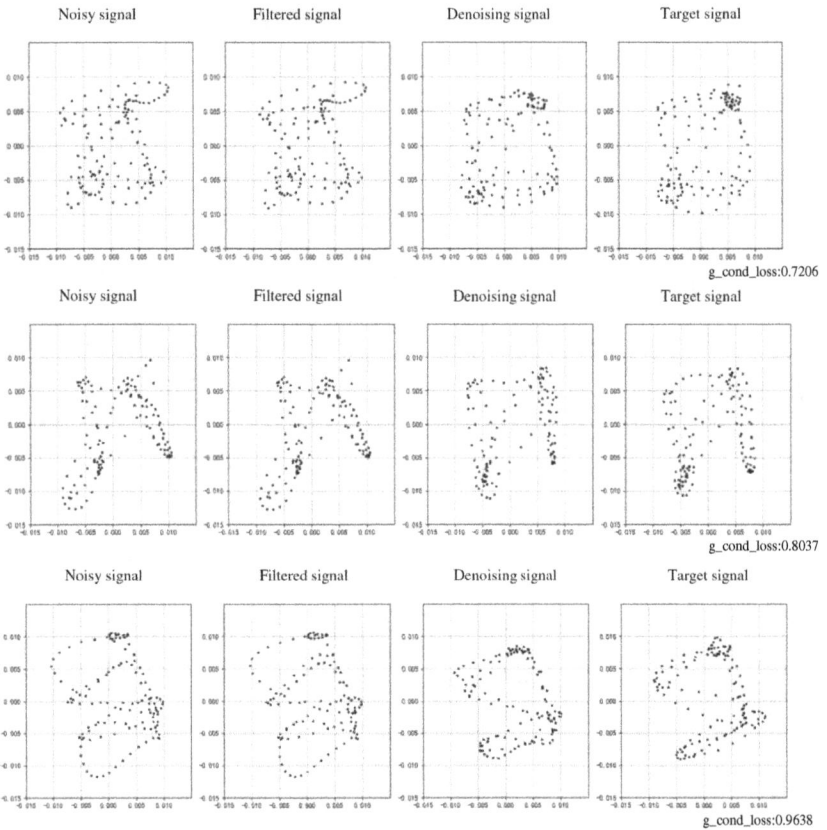

Fig. 9: Comparison of constellation diagrams for superposition.

(1) The filtered signal basically cannot reflect the denoising effect, but the denoising signal reflects the denoising effect; and

(2) The denoising signal of the overall constellation diagram is more similar to the target signal than the filtered signal.

It can be seen from Fig. 10 that the low-frequency domain denoising signal exhibits better denoising performance than the low-pass filter, and the peak value is reduced by about half. The denoising signal is closer to the target signal in the frequency domain, while the low-pass filter basically cannot reflect the denoising effect in the frequency domain.

The pure index line chart of different signals under different interferences is shown in Fig. 11. The purity index of the denoising signal

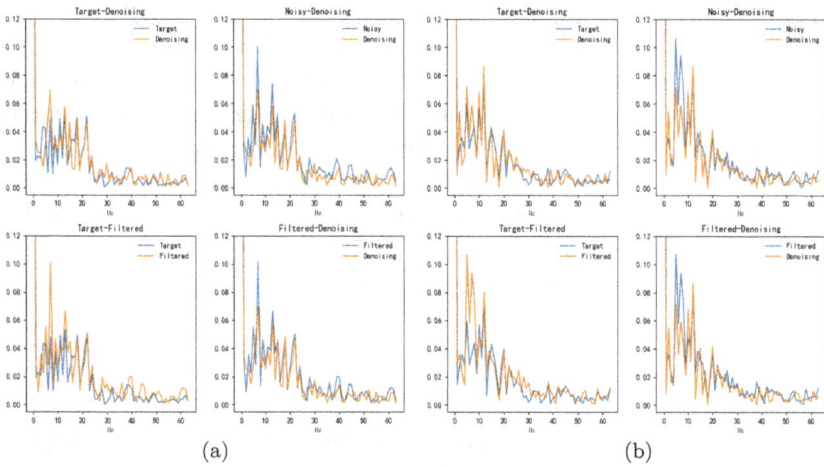

Fig. 10:   Time–frequency domain comparison for superposition.

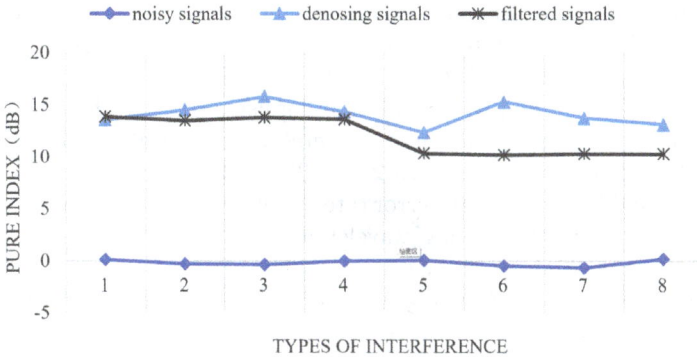

Fig. 11: Line chart of comparison of purity indicators under different interferences.

is improved by about 2.12 dB on average than the purity index of the filtered signal. In the case of interference 1, the signal purity index after filtering is about 0.3 dB higher than the denoising signal purity index, while in the other seven interference conditions, the noise signal purity index is higher than the signal purity index after filtering, especially in the case of interference 6. It is about 5 dB higher. Therefore, it can be proved that in these eight interference situations, the denoising model is better than the low-pass filter in improving the purity index.

5. Conclusion

In this chapter, a generative countermeasure network is used to carry out denoising research on radio modulation signals. The research results show that the denoising technology based on deep learning has a greater improvement than the traditional low-pass filter. The denoising model adopts end-to-end results and training. After the end, there is no need to adjust the parameters according to the prior knowledge of the signal, and the applicability is stronger.

References

1. S. D. Stearns and D. R. Hush, *Digital Signal Processing with Examples in MATLAB®*. Boca Raton, Fla: CRC Press, 2016.
2. A. Boudraa, J. Cexus, and Z. Saidi, Emd-based signal noise reduction, *International Journal of Signal Processing.* **1**(1), 33–37 (2004).
3. G. Kaushal, A. Singh, and V. Jain, Better approach for denoising EEG signals, in *2016 5th International Conference on Wireless Networks and Embedded Systems (WECON)*, 2016, pp. 1–3.
4. J. Joy, S. Peter, and N. John, Denoising using soft thresholding, *International Journal of Advanced Research in Electrical, Electronics and Instrumentation Engineering.* **2**(3), 1027–1032 (2013).
5. Y. Ge and D. G. Daut, Bit error rate analysis of digital communications signal demodulation using wavelet denoising, in *34th IEEE Sarnoff Symposium*, 2011, pp. 1–6.
6. J. Li, J. Wang, and Z. Xiong, Wavelet-based stacked denoising autoencoders for cell phone base station user number prediction, in *2016 IEEE International Conference on Internet of Things (iThings) and IEEE Green Computing and Communications (GreenCom) and IEEE Cyber, Physical and Social Computing (CPSCom) and IEEE Smart Data (SmartData)*, 2016, pp. 833–838.
7. Y. Kopsinis and S. McLaughlin, Development of EMD-based denoising methods inspired by wavelet thresholding, *IEEE Transactions on Signal Processing.* **57**(4), 1351–1362 (2009).
8. N. E. Huang, Z. Shen, S. R. Long, M. C. Wu, H. H. Shih, Q. Zheng, N.-C. Yen, C. C. Tung, and H. H. Liu, The empirical mode decomposition and the Hilbert spectrum for nonlinear and non-stationary time series analysis, in *Proceedings of the Royal Society of London. Series A: Mathematical, Physical and Engineering Sciences.* **454**(1971), 903–995 (1998).
9. N. E. Huang, *Hilbert-Huang Transform and Its Applications*, 2014, vol. 16, Singapore.

10. F. Agostinelli, M. R. Anderson, and H. Lee, Adaptive multi-column deep neural networks with application to robust image denoising, *Advances in Neural Information Processing Systems.* **26**, 1493–1501 (2013).
11. K. Zhang, W. Zuo, Y. Chen, D. Meng, and L. Zhang, Beyond a Gaussian denoiser: Residual learning of deep CNN for image denoising, *IEEE Transactions on Image Processing.* **26**(7), 3142–3155 (2017).
12. P. Liu, H. Zhang, K. Zhang, L. Lin, and W. Zuo, Multi-level wavelet-CNN for image restoration, in *Proceedings of the IEEE Conference on Computer Vision and Pattern Recognition Workshops*, 2018, pp. 773–782.
13. A. Azarang and N. Kehtarnavaz, A review of multi-objective deep learning speech denoising methods, *Speech Communication.* **122**, 1–10 (2020).
14. H. Y. Kim, J. W. Yoon, S. J. Cheon, W. H. Kang, and N. S. Kim, A multi-resolution approach to GAN-based speech enhancement, *Applied Sciences.* **11**(2), 721 (2021).
15. F. G. Germain, Q. Chen, and V. Koltun, Speech denoising with deep feature losses, arXiv preprint arXiv:1806.10522 (2018).
16. S. Ye, X. Hu, and X. Xu, Tdcgan: Temporal dilated convolutional generative adversarial network for end-to-end speech enhancement, arXiv preprint arXiv:2008.07787 (2020).
17. A. v. d. Oord, S. Dieleman, H. Zen, K. Simonyan, O. Vinyals, A. Graves, N. Kalchbrenner, A. Senior, and K. Kavukcuoglu, Wavenet: A generative model for raw audio, arXiv preprint arXiv:1609.03499 (2016).
18. S. Yu, J. Ma, and W. Wang, Deep learning for denoising Deep learning for denoising, *Geophysics.* **84**(6), V333–V350 (2019).
19. W. Zhu, S. M. Mousavi, and G. C. Beroza, Seismic signal denoising and decomposition using deep neural networks, *IEEE Transactions on Geoscience and Remote Sensing.* **57**(11), 9476–9488 (2019).
20. L. Casas, A. Klimmek, N. Navab, and V. Belagiannis, Adversarial signal denoising with encoder-decoder networks, in *2020 28th European Signal Processing Conference (EUSIPCO)*, 2021, pp. 1467–1471.
21. K. Antczak, Deep recurrent neural networks for ECG signal denoising, arXiv preprint arXiv:1807.11551 (2018).
22. C. Arsene, Complex deep learning models for denoising of human heart ECG signals, arXiv preprint arXiv:1908.10417 (2019).
23. W. Wei and E. Huerta, Gravitational wave denoising of binary black hole mergers with deep learning, *Physics Letters B.* **800**, 135081 (2020).
24. Y. Wang, L. Tu, J. Guo, and Z. Wang. Residual learning based RF signal denoising, in *2018 IEEE International Conference on Applied System Invention (ICASI)*, 2018, pp. 15–18.

Chapter 8

A Graph Neural Network Modulation Recognition Framework Based on Local Limited Penetrable Visibility Graph

Jinchao Zhou, Kunfeng Qiu, Zhuangzhi Chen,
Shilian Zheng, and Qi Xuan*

*Institute of Cyberspace Security,
Zhejiang University of Technology,
Hangzhou, P.R. China*

**xuanqi@zjut.edu.cn*

Modulation recognition is an important task for radio communication, and is the basis of demodulating the signal for further signal processing. Besides this, some visibility graph (VG) methods have been proposed to map time series into graphs and show reasonable representation capacity. However, the existing VG methods are not suitable for radio signals and powerful graph learning models haven't been used for multi-channel graph classification. In this work, a new modulation recognition framework based on graph neural network (GNN) is proposed. First, an improved VG method named local limited penetrable visibility graph (LLPVG) is proposed for representing radio signals. Second, a universal multi-channel graph feature extraction GNN model is proposed for corresponding I/Q signal graphs. Finally, we carry out experiments on the radio signal common dataset to prove the powerful representation ability of the proposed method, and the results show that our modulation recognition framework yields much better performance compared to typical VG methods and other radio signal classification models.

1.   Introduction

Time series are popular in the world, and I/Q radio signals as typical and essential time series in communication have attracted widespread attention recently. Besides this, the successful application of deep learning on Euclidean data has led to its rapid development on non-Euclidean data, i.e., graphs. The application of mature complex network theory and the powerful feature extraction capabilities of graph neural networks (GNNs) led to the idea of the combination of I/Q radio signals and GNNs.

In the field of communication, the purpose of modulation recognition is to solve the problem of determining what the signal is, after which other signal processing can proceed smoothly. The traditional machine learning modulation recognition is usually completed in two steps: feature extraction and classification. Soliman et al.[1] and Subasias et al.[2] use Fourier transform[3] and wavelet transform[4] to preprocess the signal, and then extract the ordered cyclic spectrum, high-order cumulant, cyclostationary characteristics, power spectrum,[5] and other characteristics of the signal. Based on these features, traditional classification methods in machine learning, such as decision trees,[6] random forests,[7] and support vector machines (SVMs),[8] can be used to classify time series.

As a branch of machine learning, deep learning is a combination of powerful automatic feature extractors and efficient classifiers. It replaces the process of selecting handcrafted features and effectively uses the existing features to complete classification tasks. Modern machine learning methods such as deep learning have begun to increasingly develop rapidly in the field of radio communications. Generally, there are two main ways of achieving I/Q radio signal classification in deep learning. The first one is taking signals directly as the input to train an improved recurrent neural network (RNN) classification model. For instance, Hochreiter and Schmidhuber[9] proposed a long short-term memory network (LSTM) framework based on RNN, which is designed to retain the time-related features and deepen the RNN model to capture high-level features. Although LSTM can obtain satisfactory accuracy on time series classification tasks and the model is very small, it takes a really long time to train. With the rapid development of convolution neural networks (CNNs), mapping signals into images and then utilizing 1DCNN[10] or 2DCNN[11–13] comes to be the second quick and effective settlement, though the model is hundreds of times larger than LSTM. Since the CNN-based methods have been applied successfully in modulation recognition and the GNN[14–16] model has shown its powerful node feature and structure learning ability though the model is relatively simple, we consider mapping I/Q radio signals to graphs and then utilizing GNN can mine more sequence-structure characteristics of signals and the model size is close to LSTM at the same time. So, the first thing is to make a bridge between signals and graphs.

Visibility graph (VG)[17] is considered a feasible and successful model for mapping time series to graph, one that shows the ability to characterize different types of time series, such as period, chaos, and fractal signals. Since VG was first proposed in 2008, a large number of VG model variants have emerged to establish reasonable graphs for time series in different fields. For instance, Lacasa *et al.*[18] proposed the horizontal visibility graph (HVG), which is relatively simple but efficient in EEG signal classification tasks. Gao *et al.*[19] proposed limited penetrable visibility graph (LPVG), the aim of which is to reconstruct the edges covered by noise in VG, and applied the robust model in typical three-phase flow pattern recognition successfully. Simply put, different variants change the visibility mapping rules, and these VG variants have been widely used to process all kinds of tasks in finance,[20–22] physiology,[23,24] meteorology and oceanography,[25] geography,[26] etc. However, the mentioned mapping models do not perform well in the task of radio signal modulation classification due to differences between the modulation symbols being quite large and inevitably noisy. Therefore, an improved VG model is needed for modulation signals to capture the features in each modulation symbol while the time-related features can still be retained.

After constructing graphs from I/Q radio signals reasonably, the modulation recognition task changes to graph classification task naturally. At present, the advanced methods used in graph theory mainly include graph embedding and GNN, both of which are designed to form a representation of the entire graph. Graph embedding is friendly to traditional machine learning methods, as well as for extracting some handcrafted attributions of graphs, such as the number of nodes, the number of edges, diameter, the clustering coefficient, etc., or using some unsupervised learning algorithms[27] to automatically extract features. Further, since the architecture of the GNN was designed, GNN has shown its powerful capability in graph tasks. GNN can embed nodes according to their neighbor information, that is, aggregate and optimize node features. In the past few years, in order to deal with complex graph data, the generalization and definition of important operators have developed at a quick pace. In addition to graph convolutional networks (GCNs), many new GNN models have emerged, such as GraphSAGE,[15] Diffpool,[16] GAT,[28] etc. And all these GNN models have many applications across fields, such as computer vision,[29–31] traffic,[32–34] chemistry,[35–38] etc. However, there are almost no applications that combine modulation signals with GNN to realize the automatic classification algorithm.

In this chapter, to realize modulation recognition utilizing the powerful feature extraction capability of GNNs, and simultaneously match the performance to CNNs, first we introduce a visibility slider to the LPVG models to be suitable for I/Q radio signals, expand underlying features based on communication knowledge, and design an I/Q radio signal classification framework with GNN. To validate the effectiveness of the proposed framework, first we test the improved VG model with the original VG model in a similar traditional machine learning framework and explore the impact of different lengths of the visibility slider on classification accuracy. Furthermore, the classification performance and model size of our proposed GNN deep learning framework are tested compared to LSTM and some CNN models.

The main contributions of this chapter are as follows:

- We improved the LPVG model, namely local limited penetrable visibility graph (LLPVG), by introducing a visibility slider into LPVG, to make the VG model focus on the local structural feature. LLPVG model reduces the connection between the current sampling point and the point far away, which used to be noise that interfered with model training. Further, the presence of the visibility slider greatly reduces the graph construction time apparently.
- For IQ radio signals, we extract the amplitude A and phase W information as two new channels, according to the communication knowledge. The dual-channel IQ signals are expanded to the four-channel $IQAW$ signal for constructing multi-channel graphs.
- We develop a multi-channel radio signal classification framework with GNNs for the first time. We find our modulation framework

performs much better than the advanced graph kernel machine learning, traditional RNN, and CNN framework, while the size of the model parameters is relatively much smaller.

The rest of this chapter is organized as follows. In Section 2, we briefly describe the related work on modulation recognition and typical VG methods, and then introduce the GNN. In Section 3, we introduce our multi-channel signal classification framework and LLPVG methods in detail. In Section 4, we give the experiments to validate the effectiveness of our two methods. Finally, we give the conclusion in Section 5.

2.   Related Work

In this section, some necessary background information is provided on modulation recognition, base VG models, and graph neural networks.

2.1.   *Modulation recognition*

In communication, all radio signals need to be modulated before transmission. Therefore, modulation recognition is a necessary task for demodulation and signal analysis. Automatic modulation classification (AMC)[39] is an essential technique between signal delivery and demodulation. Since the development of machine learning, AMC with deep learning models[9–11] performs quite excellently for a receiver that has limited knowledge of received signals. The aim of modulation recognition is to learn a function $f(\cdot)$ from modulation signal samples to recognize the modulation types of unseen signals. Given a signal sample S of length n:

$$S = \{x_1, x_2, \ldots, x_k, \ldots, x_n; y_{\text{true}}\}, \tag{1}$$

where x_k represents the kth sampling point's value and y_{true} is the true label or modulation type of the signal S. Modulation recognition can be defined as

$$y = f(S), \tag{2}$$

where y represents the predicting label or modulation type of the signal S.

2.2. *Visibility graph*

VG used to be a classic method in robot motion and computational geometry, and the basic idea of it is that every time point is a node in graphs and the links between two time points are connected when the two points satisfy the visibility manner. Experiments show that different types of clear time series have obviously different shapes or graph features of complex networks, and the different forms of graphs are closely related to the characteristics of time series. For instance, the degree distributions of graphs mapped from period signals exhibit regular spikes while those of graphs constructed from chaotic signals exhibit irregular and multimodal spikes. Since the VG model can construct graphs inheriting important properties of sequences, analyzing time series and the complex systems behind graphs has become popular gradually. And a series of VG variants, in simple different visibility rules, are proposed for constructing time series in various fields. Among them, LPVG is a useful and effective improved model. Gao *et al.* note that the nature of VG models is easily affected by noise while noise is common in the signals and cannot be avoided in practice. Therefore, they set a threshold that allows some local invisible points in VG to be connected to better resist the noise and be applied successfully in two-phase flow recognition.[19] Besides this, there are many other VG models, however, most of these extensions are designed for periodic and chaotic signals and are invalid when applied to modulation signals due to the weak correlation between different modulation symbols in the same modulation type.

In graph theory, a graph can be described as follows:

$$G(V, E, X), \qquad (3)$$

where $V = \{v_1, v_2, \ldots, v_N\}$ is a finite set of nodes; $N = |V|$ is the number of nodes; $E = \{e_{i,j} \mid i, j \in N\}$ is the set of edges; $e_{i,j} \neq e_{j,i}$; $X \in \mathbb{R}^{N \times F}$ is a node feature matrix and F is the node feature dimension. Thus, the VG can be formulated as

$$G = VG(S), \qquad (4)$$

where VG represents a specified model of the VG; all sampling time points $\{1, 2, \ldots, n\}$ in signal S are considered as the node set $V = \{v_1, v_2, \ldots, v_n\}$ and the sampling value is set to node feature matrix $X = \{x_1, x_2, \ldots, x_n\}^T \in \mathbb{R}^{n \times 1}$. Further, edges are established

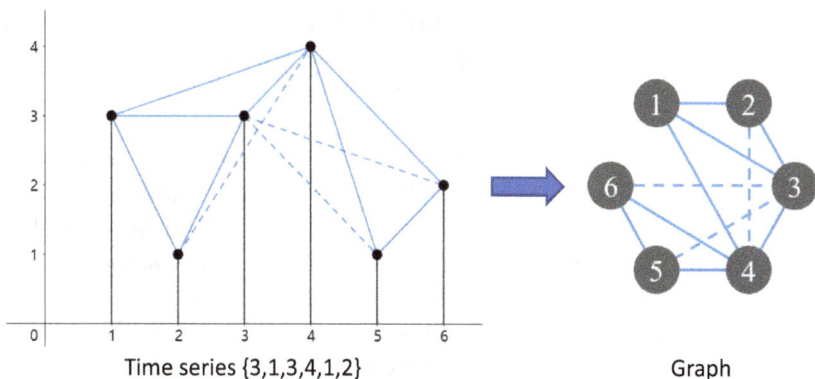

Fig. 1: Schematic diagrams of LPVG ($M = 1$), and the dashed line is the penetrated line.

according to the corresponding visibility rule. For LPVG's visibility rule, LPVG defines a limited penetrable visibility distance M additionally compared with VG (Fig. 1). Note that when $M = 0$, LPVG is the same as VG. And given two sampling points (i, x_i) and (j, x_j), then if and only if there are no more than M sampling points (k, x_k) with $i < k < j$ satisfying the following criterion:

$$x_k - x_i > (x_j - x_i)(k - i)/(j - i), \tag{5}$$

while all the other sampling points satisfy

$$x_k - x_i \leq (x_j - x_i)(k - i)/(j - i), \tag{6}$$

then, the mutual connection $e_{i,j}$ can be established between v_i and v_j.

2.3. *Graph neural network*

Driven by deep learning, researchers draw on the ideas of CNN, LSTM and deep autoencoder (AE) to design the architecture of GNNs. These early researches used recurrent neural structures to propagate neighbor information iteratively until a stable state was reached to learn the representation of the target node. In recent years, many GNN variants have been proposed with different manners of node updating. For instance, after defining the Laplacian operator and Fourier transform in the graph, graph convolution network (GCN)[14] can aggregate nodes features in frequency domain successful. Furthermore, GrarphSAGE,[15] as a typical spatial-based GNN, updates nodes by aggregating the sampled neighborhood, which solves the problem that GCN needs to know the entire graph information and be applied effectively in inductive learning. Then the new graphs learned by GNN can be further processed according to different tasks, such as node classification,[40] link prediction,[41] graph classification,[42] etc. Specifically, for graph classification tasks related to this chapter, a simple global pooling layer or a more sophisticated graph-level pooling structure such as Diffpool[16] is needed to read out the vector representation of the entire graph after convolution layers. Generally, the graph classification tasks process single-channel graph data, while IQ radio signals are multi-channel data, which means a signal is represented by multiple graphs. So the existing GNN framework needs some improvement to realize the multi-channel graph classification.

3. Methods

Our GNN modulation recognition framework is mainly composed of time-domain feature expansion, graph mapping, and GNN feature extraction. In this section, the details of these three parts are introduced. The architecture of the proposed framework is shown in Fig. 2.

3.1. *Time-domain feature expansion*

For reliable transmission and resource conservation of radio signal transmission, a modulated signal is usually converted into a pair of orthogonal IQ signals and transmitted separately. I and Q are the two sinusoids that have the same frequency and are $90°$ out of phase. Simply the modulation signal can be denoted by $S = I + jQ$. The amplitude A and phase W are also crucial features, which can be easily obtained from I and Q as follows:

$$A = \sqrt{I^2 + Q^2}, \tag{7}$$

$$W = \arctan \frac{Q}{I}, \tag{8}$$

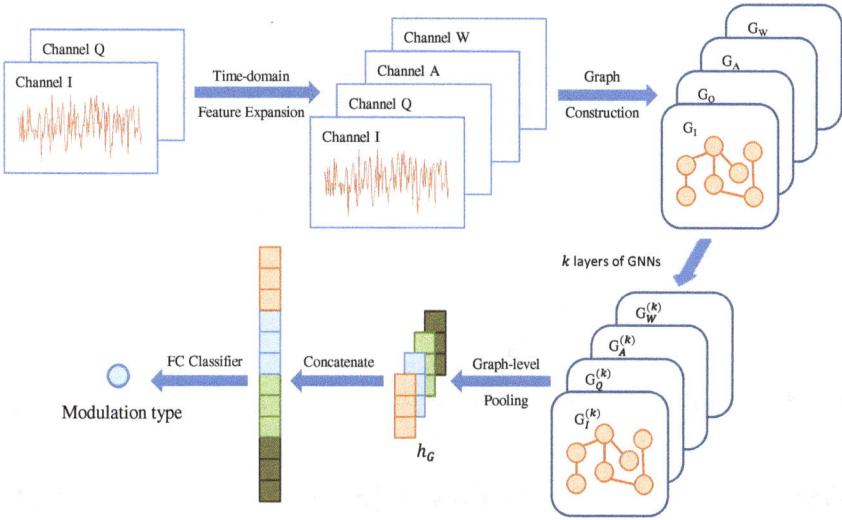

Fig. 2:   The architecture of our method.

where $W \in [0, 2\pi]$. And then dual-channel IQ signals are expanded to the four-channel $IQAW$ signal in time domain.

3.2.   *Local limited penetrable visibility graph*

In graph classification tasks, global information can sometimes help identify the category of the graph, but local information usually plays a key role. Though the VG methods can obtain both local and global features, the global features sometimes are affected by interference. For modulation signals, even with the same modulation type, the modulation symbols are quite different from each other. Therefore, global features of modulation signals may easily introduce noise to affect correct classification. A modulation symbol is composed of several sampling points, and then we consider that retaining the features of the current modulation symbol and the features of the relationship with nearby modulation symbols may be an effective way for modulation recognition.

To capture the aforementioned local features and have robustness in resisting noise, we introduce a visibility slider into the LPVG. Compared with LPVG, LLPVG only modifies the node feature matrix X and adds one more restriction in visibility criterion.

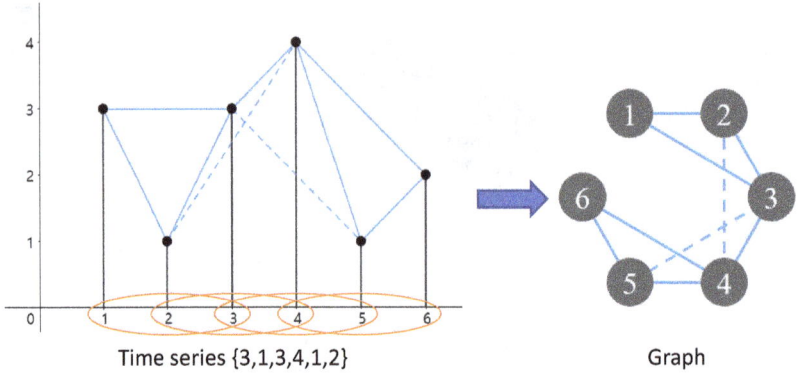

Fig. 3: Schematic diagrams of LLPVG ($M = 1$, $w = 2$), and the dashed line is the penetrated line.

Given an m-channel signal set $S_{\text{set}} = \{S_1, S_2, \ldots, S_k, \ldots, S_{m-1}, S_m\}$, limited penetrable distance M and length of visibility slider w, the graph is constructed, respectively. Here we regard S in 1 as each of the channel signals simply. For each channel signal $S = \{x_1, x_2, \ldots, x_k, \ldots, x_n; y_{\text{true}}\}$, after getting the same node set $V = \{v_1, v_2, \ldots, v_n\}$ as the LPVG, the visibility rule of LLPVG can be described such that if and only if there are no more than M sampling points (k, x_k) with $i < k < j < i + w$ satisfying Eq. (5), while all the other sampling points satisfy Eq. (6). Then, the mutual connection $e_{i,j}$ can be established between v_i and v_j. The graph construction through LLPVG is shown in Fig. 3 and the pseudocode of LLPVG is shown in Algorithms 1. The inputs of LLPVG are signal S, limited penetrable distance M, and length of visibility slider w. And the output is the Graph $G(V, E)$. As for the node feature matrix, for one-channel signals, the node feature matrix X is the same as that of LPVG. For multi-channel signals, we merge the feature of each channel as the initial node feature matrix of each channel:

$$
X = \left\{
\begin{matrix}
x_1^1 & x_1^2 & \cdots & x_1^{n-1} & x_1^n \\
x_2^1 & \ddots & \ddots & \ddots & x_2^n \\
\vdots & \ddots & x_i^j & \ddots & \vdots \\
x_{m-1}^1 & \ddots & \ddots & \ddots & x_{m-1}^n \\
x_m^1 & x_m^2 & \cdots & x_m^{n-1} & x_m^n
\end{matrix}
\right\}^T \in \mathbb{R}^{n \times m}, \tag{9}
$$

where x_i^j represents the jth sampling value of the i-channel signal, then the graph can be described as $G(V, E, X)$.

Algorithm 1 Map Time Series to Graph through LLPVG

Input: The signal sample S as shown in Eq. (1), length n of S, limited penetrable distance M, length of visibility slider w.
Output: Graph $G(V, E)$ formed by LLPVG.

1 Set the value of the hyperparameter M and w
    for $i=1$ to n-1 **do**
2   Append node i to V
     $limit = 0$
     for $j=i+1$ to $i+w$ **do**
3   **for** $k=i$ to j **do**
4          $limit = limit + 1$ if x_k satisfies Eq. (5)

5   Append edge (i, j) to E if $limit \leq M$

6 Append node n to V
    return $G(V, E)$

3.3. *Graph neural networks*

After mapping modulation signals to graphs, modulation recognition turns to graph classification. And then graph embedding and GNNs methods can be applied to extract features. In the proposed framework, we use GNN to capture the global features, which utilizes the graph structure and node features. Most common GNN models can be summarized as the same process, and we introduce the process in this subsection.

For graph classification, given a set of graphs $G_{\text{set}} = \{G_1, \ldots, G_N\}$ and their labels $y_{\text{set}} = \{y_1, \ldots, y_N\}$, let $G(V, E, X)$ in Eq. (3) denote a graph with node feature matrix $X = \{X_v\}$ for $v \in V$. We need to learn the graph representation methods to represent each graph G as a feature vector h_G, which helps predict the label of the graph G as follows:

$$h_G = GNNs(G). \tag{10}$$

GNN models for graph classification can be divided into three steps, i.e., aggregate, update, and readout. First, a kth layer GNN aggregate neighbors node features within kth hop:

$$a_v^{(k)} = \text{Aggregate}^{(k)} \left(\left\{ h_u^{(k-1)} : u \in N(v) \right\} \right), \quad (11)$$

where $N(v)$ is the neighborhoods of the node v, $h_u^{(k-1)}$ is the node u's feature vector of the previous update layer. As for Aggregate$(\cdot)$, it can be summation, averaging, weighted summation, etc. For example, the Aggregate$(\cdot)$ in GraphSAGE can be formulated as

$$a_v^{(k)} = \text{MAX} \left(\left\{ \text{ReLU} \left(W \cdot h_u^{(k-1)} \right), \forall u \in N(v) \right\} \right), \quad (12)$$

where W is a learnable weight matrix and MAX$(\cdot)$ represents a node-level max-pooling. After aggregating neighborhood features, the node feature vector can be updated as

$$h_v^{(k)} = \text{Update}^{(k)} \left(h_v^{(k-1)}, a_v^{(k)} \right), \quad (13)$$

where $h_v^{(k)}$ represents the kth layer feature vector of node v. As for Update$(\cdot)$, it can be summation, averaging, or linear mapping after concatenation. For example, one kind of Update in GraphSAGE could be formulated as

$$h_v^{(k)} = W \cdot \left[h_v^{(k-1)}, a_v^{(k)} \right]. \quad (14)$$

Finally, a Readout layer is applied to obtain the entire graph's representation h_G by aggregating all node features from the final layer:

$$h_G = \text{Readout} \left(\left\{ h_v^{(K)} \mid v \in G \right\} \right). \quad (15)$$

The Readout$(\cdot)$ could be summation, averaging, or hierarchical graph-level pooling.

4. Experiment

In this section, we mainly introduce the specific implementation processes and the results of LLPVG, and compare with those obtained by VG and LPVG on the public radio dataset, to validate the effectiveness of our methods.

4.1.   Datasets

The dataset used in the experiments is an open synthetic radio modulation dataset RML2016.10a.[11] It is a high-quality radio signal simulation dataset generated by GNU Radio, which was first released at the 6th GNU Annual Radio Conference. This signal dataset contains 11 modulation types (8 digital and 3 analog). Digital modulations include BPSK, QPSK, 8PSK, 16QAM, 64QAM, BFSK, CPFSK, and PAM4. Analog modulations include WB-FM, AM-SSB, and AM-DSB. Each modulation type contains 20 different signal-to-noise ratios (SNRs), and each SNR contains 1,000 samples. Each sample consists of an in-phase signal I and a quadrature signal Q. And each signal contains 128 sampling points. So the size of the complete dataset is $220{,}000 \times 2 \times 128$. In the experiment, we divide the training set and the test set in a ratio of 4:1. Considering the balance of the SNR signal samples, we randomly select 80% samples of each SNR in each modulation type as the training set, and the rest as the test set.

4.2.   Baselines

In this subsection, the proposed model is compared with the following models: (1) gated recurrent unit network (GRU),[43] (2) LSTM recurrent neural network,[9] (3) DCNN in Ref. 10, (4) 2DCNN in Ref. 44, (5) limited penetrable visibility graph (LPVG-G2V) ($M = 1$) with Graph2vec, and (6) limited penetrable visibility graph (LPVG-GraphSAGE) ($M = 1$) with GraphSAGE.

4.3.   The experimental settings

For LLPVG ($M = 1$), we first make a simple comparison for the different visibility slider w values in $\{4, 8, 16\}$. The experiment uses a graph embedding method Graph2vec to automatically extract the graph-level features and a Random Forest (RF) as the classifier. And then, we choose the best performing LLPVG with $w = 4$ for comparison with the aforementioned baselines.

All the experiments are run on NVIDIA Tesla V100 based on the PyTorch[45] deep learning framework for CNN and RNN models and PyTorch Geometric[46] framework for GNN models. And all the baselines and our proposed model have the same general hyper-parameters, such as batch size, training epoch, and learning rate. Batch size, epoch, and learning rate are set to 128, 100, and 0.001, respectively. For RNN models, layers of both GRU and LSTM are set to 2 and hidden units are set to 128. For CNN models, 1DCNN is composed of six residual stack units and two fully connected layers, and 2DCNN is composed of two convolution layers and two fully connected layers. For GNN model, we use a three layer GraphSAGE and two fully connected layers for comparison, and the hidden feature is set to 64. For VG models, limited penetrable distance M is set to 1 in both LPVG and LLPVG.

4.4. *Results and discussion*

We first do a preliminary exploration under the framework of machine learning. The feature extraction method is graph2vec and the classifier is Random Forest. The results of LLPVG ($M = 1, w = \{4, 8, 16\}$) and LPVG ($M = 1$) under the same framework are shown in Table 1. We show the classification accuracy of signals with different SNRs. According to the information of the dataset, a modulation symbol consists of 8 sampling points, and the results show that when $w = 4$, the classification accuracy is highest in most SNRs. When $w = 4$, the range of connections that can be established at each sampling point is just within the length of one modulation symbol. As w increases, the classification performance gradually decreases. It demonstrates that the connection between different modulation symbols will introduce noise and local information is more important in this classification task.

Then we choose the best performing LLPVG with $w = 4$ for comparison with the aforementioned baselines, and a three-layer Graph-SAGE is used as the GNN model for both LPVG and LLPVG. The results are shown in Table 2, and we also present the accuracy of different SNR signals. In the real world, the SNR of radio signals is generally higher than -6 dB, and the results show that our proposed model outperforms other baselines, especially for the high SNR signals. Besides, we find that traditional machine learning framework performs better than deep learning models in classifying the signals of very low SNR. Low SNR signals have more noise interference. With continuous iterations, deep learning models will learn more high SNR

Table 1:   The accuracy of different SNR signals under the framework of Graph2vec and Random Forest.

SNR (dB)	LLPVG-G2V			Baseline LPVG-G2V (%)
	$w = 4$ (%)	$w = 8$ (%)	$w = 16$ (%)	
18	**87.91**	87.05	86.91	84.77
16	**87.86**	85.14	85.91	84.95
14	**86.36**	83.95	84.55	83.59
12	**86.95**	85.41	85.45	83.27
10	**87.55**	86.32	85.95	83.86
8	**87.36**	84.77	83.95	82.14
6	**85.50**	83.00	83.64	79.27
4	**83.23**	80.91	80.86	76.82
2	**75.59**	73.59	72.23	67.91
0	**65.77**	64.27	62.27	60.86
−2	**52.23**	51.86	50.68	47.14
−4	**47.18**	45.05	44.59	43.23
−6	41.91	**42.05**	38.27	38.50
−8	**35.45**	31.32	31.77	32.55
−10	**27.64**	25.45	24.00	24.68
−12	22.27	**22.82**	20.91	22.73
−14	**21.09**	19.36	19.59	19.91
−16	**19.45**	19.36	18.32	18.86
−18	18.55	18.95	19.14	**19.36**
−20	**19.00**	18.55	18.82	18.86
>5	**87.07**	85.09	85.19	83.12
≥0	**83.41**	81.44	81.17	78.75
>−5	**77.79**	75.94	75.58	73.15
Total	**56.94**	55.46	54.89	53.66

Note: The bold entries represent the best values.

signal features to improve overall accuracy. And traditional machine learning models extract some specific features, while these retained features can help distinguish the low SNR signals. Whether this specific feature can be combined with a deep learning model is worth further exploration.

Finally, we present the model size of each deep learning model as shown in Table 3. All the models show reasonable classification accuracy, but the model size is quite different. The proposed model is the second smallest in these deep learning models, while the accuracy is the highest. In practical applications, the smaller the model,

Table 2: The accuracy of different SNR signals under the framework of deep learning.

dB	GRU (%)	LSTM (%)	1DCNN (%)	2DCNN (%)	LPVG-GraphSAGE (%)	**LLVPG-GraphSage (%)**
18	85.82	88.91	88.77	79.09	88.14	**91.32**
16	87.00	90.23	89.59	78.91	89.27	**91.23**
14	86.86	89.50	90.05	79.68	88.68	**90.77**
12	85.86	89.50	89.68	79.14	88.36	**90.18**
10	87.32	90.55	91.14	78.95	88.91	**91.45**
8	86.41	89.82	89.68	78.50	87.45	**90.27**
6	86.09	89.64	89.45	79.05	88.14	**90.32**
4	85.32	88.77	88.91	78.77	86.64	**89.32**
2	83.32	87.73	87.50	77.68	85.82	**87.91**
0	82.36	84.86	85.77	76.55	81.55	**87.64**
−2	74.50	75.95	76.73	70.00	74.77	**81.27**
−4	62.41	66.50	65.41	60.36	63.73	**70.59**
−6	48.91	51.32	49.23	45.14	50.00	**52.36**
−8	37.32	**39.41**	34.77	29.95	35.18	37.59
−10	24.41	**25.73**	24.18	17.18	23.45	24.41
−12	16.14	15.55	**16.77**	12.36	14.95	15.50
−14	**14.27**	11.86	12.95	10.91	11.68	12.00
−16	10.14	9.73	**10.18**	9.27	9.18	9.36
−18	9.45	9.82	9.55	9.64	9.09	**9.86**
−20	**10.05**	9.77	10.00	9.55	9.73	9.50
>5	86.48	89.73	89.77	79.05	88.42	**90.79**
≥0	85.64	88.95	89.05	78.63	87.30	**90.04**
>−5	82.77	86.00	86.06	76.39	84.29	**87.69**
Total	58.20	60.26	60.02	53.03	58.74	**61.14**

Note: The bold entries represent the best values.

Table 3: The model size of each deep learning model.

Method	Accuracy (%)	Params size (MB)
GRU	58.20	2.45
LSTM	60.26	0.82
1DCNN	60.02	0.4
2DCNN	53.03	10.83
LPVG-GraphSAGE	58.74	0.55
LLVPG-GraphSAGE	61.14	0.55

the better the terminal deployment. How to balance model size and classification performance is also worthy of further exploration.

5.   Conclusion

In this chapter, an improved LLPVG is proposed for modulation radio signals. The LLPVG ignores most noisy global features and retains the connection within the length of a modulation symbol. After mapping radio signals into graphs, modulation recognition turns to graph classification. Then combined with GNN, a new modulation recognition framework in graph-domain is proposed. The experimental results demonstrate that the proposed framework outperforms other baseline models, while simultaneously the size of the model is quite small.

In the future, we will explore the factors that affect the classification accuracy of signals with different SNRs. And then we will try to design an automatic graph learning model of time series mapping to graph, which can update the graph structure and node features during iteration.

References

1. S. S. Soliman and S.-Z. Hsue, Signal classification using statistical moments, *IEEE Transactions on Communications.* **40**(5), 908–916 (1992).
2. A. Subasi, Eeg signal classification using wavelet feature extraction and a mixture of expert model, *Expert Systems with Applications.* **32**(4), 1084–1093 (2007).
3. R. N. Bracewell and R. N. Bracewell, *The Fourier Transform and Its Applications*, 1986, vol. 31999, New York: McGraw-Hill.
4. M. Antonini, M. Barlaud, P. Mathieu, and I. Daubechies, Image coding using wavelet transform, *IEEE Transactions on Image Processing.* **1**(2), 205–220 (1992).
5. B. Mollow, Power spectrum of light scattered by two-level systems, *Physical Review.* **188**(5) (1969).
6. J. R. Quinlan, Induction of decision trees, *Machine Learning.* **1**(1), 81–106 (1986).
7. A. Liaw, M. Wiener *et al.*, Classification and regression by randomforest, *R News.* **2**(3), 18–22 (2002).

8. J. A. Suykens and J. Vandewalle, Least squares support vector machine classifiers, *Neural Processing Letters.* **9**(3), 293–300 (1999).
9. S. Hochreiter and J. Schmidhuber, Long short-term memory, *Neural Computation.* **9**(8), 1735–1780 (1997).
10. T. J. O'Shea, T. Roy, and T. C. Clancy, Over-the-air deep learning based radio signal classification, *IEEE Journal of Selected Topics in Signal Processing.* **12**(1), 168–179 (2018).
11. T. J. O'Shea, J. Corgan, and T. C. Clancy, Convolutional radio modulation recognition networks, in *International Conference on Engineering Applications of Neural Networks*, 2016, pp. 213–226.
12. S. Peng, H. Jiang, H. Wang, H. Alwageed, Y. Zhou, M. M. Sebdani, and Y.-D. Yao, Modulation classification based on signal constellation diagrams and deep learning, *IEEE Transactions on Neural Networks and Learning Systems.* **30**(3), 718–727 (2019). doi: 10.1109/TNNLS. 2018.2850703.
13. Y. Lin, Y. Tu, Z. Dou, and Z. Wu, The application of deep learning in communication signal modulation recognition, in *2017 IEEE/CIC International Conference on Communications in China (ICCC)*, 2017, pp. 1–5. doi: 10.1109/ICCChina.2017.8330488.
14. T. N. Kipf and M. Welling, Semi-supervised classification with graph convolutional networks, arXiv preprint arXiv:1609.02907 (2016).
15. W. L. Hamilton, R. Ying, and J. Leskovec. Inductive representation learning on large graphs, in *Proceedings of the 31st International Conference on Neural Information Processing Systems*, 2017, pp. 1025–1035.
16. R. Ying, J. You, C. Morris, X. Ren, W. L. Hamilton, and J. Leskovec, Hierarchical graph representation learning with differentiable pooling, arXiv preprint arXiv:1806.08804 (2018).
17. L. Lacasa, B. Luque, F. Ballesteros, J. Luque, and J. C. Nuno, From time series to complex networks: The visibility graph, *Proceedings of the National Academy of Sciences.* **105**(13), 4972–4975 (2008).
18. W.-J. Xie, R.-Q. Han, and W.-X. Zhou, Tetradic motif profiles of horizontal visibility graphs, *Communications in Nonlinear Science and Numerical Simulation.* **72**, 544–551 (2019).
19. L. Cai, J. Wang, Y. Cao, B. Deng, and C. Yang, Lpvg analysis of the eeg activity in alzheimer's disease patients, in *2016 12th World Congress on Intelligent Control and Automation (WCICA)*, 2016.
20. Y. Yang, J. Wang, H. Yang, and J. Mang, Visibility graph approach to exchange rate series, *Physica A: Statistical Mechanics and Its Applications.* **388**(20), 4431–4437 (2009).
21. M.-C. Qian, Z.-Q. Jiang, and W.-X. Zhou, Universal and nonuniversal allometric scaling behaviors in the visibility graphs of world stock

market indices, *Journal of Physics A: Mathematical and Theoretical.* **43**(33), 335002 (2010).

22. N. Wang, D. Li, and Q. Wang, Visibility graph analysis on quarterly macroeconomic series of china based on complex network theory, *Physica A: Statistical Mechanics and Its Applications.* **391**(24), 6543–6555 (2012).

23. M. Ahmadlou, H. Adeli, and A. Adeli, New diagnostic eeg markers of the alzheimer's disease using visibility graph, *Journal of Neural Transmission.* **117**(9), 1099–1109 (2010).

24. Z.-G. Shao, Network analysis of human heartbeat dynamics, *Applied Physics Letters.* **96**(7), 073703 (2010).

25. A. Charakopoulos, G. Katsouli, and T. Karakasidis, Dynamics and causalities of atmospheric and oceanic data identified by complex networks and granger causality analysis, *Physica A: Statistical Mechanics and its Applications.* **495**, 436–453 (2018).

26. J. Elsner, T. Jagger, and E. Fogarty, Visibility network of united states hurricanes, *Geophysical Research Letters.* **36**(16) (2009).

27. A. Narayanan, M. Chandramohan, R. Venkatesan, L. Chen, Y. Liu, and S. Jaiswal, graph2vec: Learning distributed representations of graphs, arXiv preprint arXiv:1707.05005 (2017).

28. P. Veličković, G. Cucurull, A. Casanova, A. Romero, P. Lio, and Y. Bengio, Graph attention networks, *arXiv preprint arXiv:1710.10903* (2017).

29. A. K. Debnath, R. L. Lopez de Compadre, G. Debnath, A. J. Shusterman, and C. Hansch, Structure-activity relationship of mutagenic aromatic and heteroaromatic nitro compounds correlation with molecular orbital energies and hydrophobicity, *Journal of Medicinal Chemistry.* **34**(2), 786–797 (1991).

30. P. D. Dobson and A. J. Doig, Distinguishing enzyme structures from non-enzymes without alignments, *Journal of Molecular Biology.* **330** (4), 771–783 (2003).

31. K. M. Borgwardt, C. S. Ong, S. Schönauer, S. Vishwanathan, A. J. Smola, and H.-P. Kriegel, Protein function prediction via graph kernels, *Bioinformatics.* **21**(suppl 1), i47–i56 (2005).

32. P. Cui, X. Wang, J. Pei, and W. Zhu, A survey on network embedding, *IEEE Transactions on Knowledge and Data Engineering.* **31**(5), 833–852 (2018).

33. A. Bojchevski, O. Shchur, D. Zügner, and S. Günnemann, Netgan: Generating graphs via random walks, in *International Conference on Machine Learning*, 2018, pp. 610–619.

34. Y. Seo, M. Defferrard, P. Vandergheynst, and X. Bresson, Structured sequence modeling with graph convolutional recurrent networks,

in *International Conference on Neural Information Processing*, 2018, pp. 362–373.

35. Q. Li, Z. Han, and X.-M. Wu, Deeper insights into graph convolutional networks for semi-supervised learning, in *Thirty-Second AAAI Conference on Artificial Intelligence*, 2018, pp. 3538–3545.

36. A. Micheli, D. Sona, and A. Sperduti, Contextual processing of structured data by recursive cascade correlation, *IEEE Transactions on Neural Networks.* **15**(6), 1396–1410 (2004).

37. A. Sperduti and A. Starita, Supervised neural networks for the classification of structures, *IEEE Transactions on Neural Networks.* **8**(3), 714–735 (1997).

38. D. Xu, Y. Zhu, C. B. Choy, and L. Fei-Fei, Scene graph generation by iterative message passing, in *Proceedings of the IEEE Conference on Computer Vision and Pattern Recognition*, 2017, pp. 5410–5419.

39. Z. Zhu and A. K. Nandi, *Automatic Modulation Classification: Principles, Algorithms and Applications*, 2015, New York, NY, USA: John Wiley & Sons.

40. O. Shchur, M. Mumme, A. Bojchevski, and S. Günnemann, Pitfalls of graph neural network evaluation, arXiv preprint arXiv:1811.05868 (2018).

41. M. Zhang and Y. Chen, Link prediction based on graph neural networks, *Advances in Neural Information Processing Systems.* **31**, 5165–5175 (2018).

42. F. Errica, M. Podda, D. Bacciu, and A. Micheli, A fair comparison of graph neural networks for graph classification, arXiv preprint arXiv:1912.09893 (2019).

43. D. Hong, Z. Zhang, and X. Xu, Automatic modulation classification using recurrent neural networks, in *2017 3rd IEEE International Conference on Computer and Communications (ICCC)*, 2017, pp. 697–700.

44. T. J. O'Shea, J. Corgan, and T. C. Clancy, Convolutional radio modulation recognition networks, in *International Conference on Engineering Applications of Neural Networks*, 2016, pp. 213–226.

45. A. Paszke, S. Gross, S. Chintala, G. Chanan, E. Yang, Z. DeVito, Z. Lin, A. Desmaison, L. Antiga, and A. Lerer, Automatic differentiation in pytorch, *In NIPS 2017 Autodiff Workshop: The Future of Gradient-based Machine Learning Software and Techniques* (2017).

46. M. Fey and J. E. Lenssen, Fast graph representation learning with PyTorch Geometric, in *ICLR Workshop on Representation Learning on Graphs and Manifolds*, 2019.

Part III
Network Applications

Chapter 9

Study of Autonomous System Business Types Based on Graph Neural Networks

Songtao Peng, Lu Zhang, Xincheng Shu, Zhongyuan Ruan, and Qi Xuan*

Institute of Cyberspace Security, Zhejiang University of Technology, Hangzhou, P.R. China

**xuanqi@zjut.edu.cn*

An accurate understanding of the characteristics of autonomous systems (AS) is related to both technical and economic aspects of internet inter-domain structure construction. Based on the Internet AS-level topology, algorithms are proposed to classify AS relationships. However, the existing tasks achieve information mining through a large amount of manual working. In this chapter, we enhance AS business-type inference by using AS relationship as a feature for the first time. Meanwhile, the proposed new framework ASGNN, considers AS static attributes, global network structure, and local link features to determine AS types. Experiments on real-world internet topological data validate the effectiveness of our approach. ASGNN outperforms a series of baselines and helps further understand the structure and evolution of the internet.

1. Introduction

The internet is composed of more than 70,000 ASes. An AS is a network unit that has the capability to independently decide which routing protocol should be used in the system. ASes of different sizes, functions, and business objectives form several *AS types* that jointly form what we know as the global internet. Routing information is exchanged between ASes through border gateway protocol (BGP). Meanwhile, routing information can be effectively used to construct the topological graph of the AS interconnection, and thus, can implement strategic decisions at the AS-level. Typically, in AS-level topology, the business relationships between connected ASes are broadly classified into (1) customer-provider (C2P), (2) peer–peer (P2P), and (3) sibling relationship (S2S). A series of studies on these relationships include topology flattening,[1,2] network congestion detection,[3–5] internet security check,[6–9] variable routing protocol based attack designing,[10] and network characteristics analysis.[11]

In this chapter, our study is motivated by the desire to better understand this complex ecosystem and the behavior of entities (ASes). We focus on the accurate knowledge of AS business relationships, which is essential for understanding both technical and economic aspects of the internet. The model research of AS classification is only in its infancy so far. There is currently no unified classification standard. For different problems, researchers propose corresponding categories and characteristics. Typically, the classification is performed based on the degree or prefixes of an AS. Meanwhile, most studies use machine learning algorithms, such as decision trees, to get the positive predictive value (PPV), which still has room for improvement.

Since the AS relationship was introduced in 2001,[12] many inference algorithms[13–20] have been proposed. Notably, it is found that some advanced algorithms, including AS-Rank,[15] ProbLink,[17] and TopoScope,[18] have a lower prediction error rate,[15] and behave surprisingly well in predicting *hard links*[17] or *hidden links*.[18] However, most of the current algorithms separate the two tasks (AS Classification and AS Relationship). The type of AS reflects its function and affects the relationship with other ASes. The AS relationship highlights the characteristics of AS from the link structure. In general, they can improve classification performance through interaction. In order to explore the real-world situation, we conduct comparative experiments to reflect the mutual influence between AS Classification and AS Relationship. By synthesizing and analyzing the important indicators in state-of-the-art inference methods, such as *triplet, distance to clique*, and so on,[15] we quantitatively explain the importance of existing features in a variety of tasks for the current AS-level topology.

Facing the situation where there is correlation between nodes and nodes, nodes and links, and links and links, we attempt to combine the construction of the network with the algorithm to improve inference performance. We develop ASGNN, which is a new graph neural network (GNN) framework for AS classification. The convolution process of ASGNN on the internet topology focuses on the static properties of AS itself and enhances the connection between AS and its neighbors. Therefore, ASGNN has the ability to summarize the training set with relevant feature information and generalize it to the testing set. In particular, the main contributions of this chapter are summarized as follows.

(1) We focus on the AS classification task for AS-level internet networks. For the AS classification problem, we introduce the AS relationship property as a new feature to the task. In terms of AS relational data acquisition, we aggregate multiple state-of-the-art algorithms based on the idea of hard voting to obtain the more comprehensive dataset, which can solve the deviation and limitation of the training data of the previous research.

(2) We study the relationship between nodes and edges in AS-level networks, and map them to the important issue of AS classification. Then we develop a new GNN framework ASGNN to improve the performance of the inference problem under various scenes, which takes into account both the global and local topological properties simultaneously. To the best of our knowledge, this is the first work to use GNN to solve this problem.

(3) Experimental results show the best performance of our ASGNN in AS classification task compared with a series of baselines.

The rest of the chapter is organized as follows. Section 2 introduces related work and describes in detail the work done in recent years on AS classification and AS relationship inference. Section 3 details the source and processing of the data. Section 4 describes the challenges of the current researches and the basic ideas for combining with the GNN model. Section 5 details the design and implementation of our framework ASGNN. In Section 6, we conduct extensive experiments. Section 7 concludes this chapter.

2. Related Work

In this part, we briefly review the background and the related works on AS Classification and AS relationship.

2.1. AS classification

Researchers have developed techniques decomposing the AS topology into different levels or tiers based on connectivity properties of BGP-derived AS graphs. Govindan et al.[21] proposed to classify ASes into four levels based on their AS degree. Ge et al.[22] classified ASes into seven tiers based on inferred customer-to-provider relationships. Their classification exploited the idea that provider ASes should be in higher tiers than their customers. Subramanian et al.[23] classified ASes into five tiers based on inferred customer-to-provider as well as peer-to-peer relationships.

Dimitropoulos et al.[24] suggested that an AS n"node" can represent a wide variety of organizations, e.g., large ISPs, small ISPs, customer ASes, Universities, internet exchange points (IXPs), and network information centers (NICs). They introduced a radically new approach *AdaBoost* based on machine learning techniques to map all the ASes in the internet into a natural AS taxonomy, and successfully classified 95.3% of ASes with expected accuracy of 78.1%.

Dhamdhere et al.[25] attempted to measure and understand the evolution of the internet ecosystem during the last 12 years. They proposed to use the machine learning method *decision tree* to classify ASes into a number of types depending on their function and business type using observable topological properties of those ASes. The AS types they considered were large transit providers, small transit providers, content/access/hosting providers, and enterprise networks. They were able to classify ASes into these AS types with an accuracy of 75–80%.

CAIDA used a ground-truth dataset from PeeringDB and trained a *decision tree* classifier using a number of features. They introduced classification features customer, provider, peer degrees, and size of customer cone in number of ASes based on the results of their algorithm AS-Rank. It considered each AS's size of the IPv4 address space advertised and number of domains from the Alexa top 1 million list hosted by the AS. They used half of the ground-truth data to validate the machine-learning classifier, and the PPV of the classifier is around 70%.

2.2. AS relationship

The Internet topology at the AS level is typically modeled using a simple graph where each node is an AS and each link represents a business relationship between two ASes. These relationships reflect who pays whom when traffic is exchanged between the ASes. They are the keys to the normal operation of the internet ecosystem. Traditionally, these relationships are categorized into (1) customer–provider (C2P), (2) peer–peer (P2P), and (3) sibling relationships (S2S).[26] However, other forms of relationships exist as well. In a C2P relationship, the customer is billed for using the provider to reach the rest of the internet. The other two types of relationships are in general settlement-free. In other words, no money is exchanged between the two parties involved in a P2P or S2S relationship.

Understanding of AS relationship is vital to the technical research and economic exchanges of the inter-domain structure of the internet. The relationship between ASes is regarded as private information by various organizations, institutions, and operators and not published on the open platform. By considering the internet as a complex network, various AS relationship inference algorithms have been proposed to predict the AS-level structural relationship of the internet, which is of particular significance for internet security.

Gao[12] first proposed to enhance the representation of the AS graph by defining multiple business relationships and put forward an assumption that valid BGP paths are *valley-free*,[27] (i.e., $[C2P/S2S]^n [P2P]^{(0,1)} [P2C/S2S]^m$, $n \geq 0$, $m \geq 0$, which means a path consists of zero or more C2P or S2S links, followed by zero or one P2P links, followed by zero or more P2C or S2S links, the shape is composed of an uphill path and a downhill path or one of the two).

It plays an important role in the later process of inference in algorithm research. Since then, a series of methods[13,14,28] have been proposed to enhance the inference performance by improving the *valley-free* feature. However, the subsequent researches proved that only considering *degree* and *valley-free* features may not be enough to infer the complex relationships of the internet.

Unlike previous approaches, AS-Rank[15] did not seek to maximize the number of *valley-free* paths but relies on three assumptions about the internet inter-domain structure: (1) an AS enters into a provider relationship to become globally reachable; (2) there exists a peering clique of ASes at the top of the hierarchy; and (3) there is no cycle of P2C links. Based on these assumptions, the features of clique, transit degree, and BGP path triplets were proposed to realize the inference. Due to its high accuracy and stability, AS-Rank has been used on CAIDA[29] until now.

ProbLink[17] was the first probabilistic AS relationship inference algorithm based on naive Bayes to reduce the error rate and overcome the challenge in inferring hard links, such as non-valley-free routing, limited visibility, and non-conventional peering practices. This approach demonstrated its practical significance in detecting route leak, inferring complex relationships, and predicting the impact of selective advertisements. TopoScope[18] further used ensemble learning and Bayesian network to reduce the observation bias, and reconstructed internet topology by discovering hidden links.

Varghese et al.[16] used AdaBoost[30] algorithm to train a model that predicts the link types in a given AS graph using two node attributes: degree and minimum distance to a *Tier-1* node. However, their choice of dataset and the setting of the experimental group were not rigorous enough. Shapira et al.[19] use natural language processing (NLP) to propose a deep learning model *BGP2VEC*. In recent years, many methods have mainly focused on the inference of the relationship P2P and P2C. Giotsas et al.[20] showed that the interconnections of ASes of a real internet are much more complex and diverse and they presented a new algorithm to infer the two most common types of complex AS relationships: hybrid and partial transit. Therefore, it is necessary to propose more proper algorithms for inferring relationships to better understand the increasingly large internet topology.

3. Datasets

In this section, we elaborate on the source of the experimental BGP paths data and introduce the standard dataset for labeled sibling relationship and internet exchange point (IXP) list. On the validation dataset, we use ground-truth data from PeeringDB to train and validate various algorithms for the AS classification problem. Our voting-based training dataset is more representative of the AS relationship problem.

3.1. *BGP paths*

We collect BGP paths from Route Views[31] and RIPE NCC,[32] the most popular projects managing route information collectors and making their data flow accessible and usable by any researcher. Currently, these two projects manage 29 and 25 collectors, respectively, with more than 1,000 VPs in total (this number is growing over time). They continuously collect most of the routing information around the world. In the experiments in this chapter, we downloaded the data flow files and extracted the BGP paths with IPv4 prefixes on 12/01/2020.

During the pre-processing stage of the data, we first remove those paths that have unassigned AS numbers (ASN) using the table provided by the internet assigned numbers authority (IANA).[33] We also sanitize the BGP paths[34] containing AS loops, i.e., the same ASN in the path can only be adjacent to each other. We compress the paths that have the same consecutive ASN (i.e., from "A B C C" to "A B C").

3.2. *S2S relationship and IXP list*

We use CAIDA's AS-to-Organization mapping dataset,[35] which applies WHOIS information available from regional and national internet registries. Mapping is available from October 2009 onwards and the new mapping is added in each quarter. Our frequency of obtaining BGP path data is similar to the updating cycle. This dataset contains the ASN and the organization it belongs to. We infer the links that its endpoints managed by the same organization as sibling relationships (S2S relationships). In the following AS relationship inference experiment, we preprocess the dataset as known.

An IXP is a physical infrastructure used by internet service providers (ISPs) and content delivery networks (CDNs) to exchange internet traffic between their ASes. An IXP can be distributed and located in numerous data centers, and a single facility can contain multiple IXPs. An AS connected to a given IXP is known as a member of that IXP. Our IXP list is derived from visiting PeeringDB[36] for networks of type *"Route Server"* and extracting the ASN. Since IXP provides a neutral shared exchange structure, and clients can exchange traffic with each other after establishing a peer–peer connection, our method also removes the BGP paths contained in IXPs like the previous work. Although not all IXPs have route servers, the number of ASes contained in the newly extracted list can be considered as the lower bound. There were 336 IXP ASes in this list on 12/01/2020.

3.3. *AS classification ground-truth dataset*

To train and validate our classification approach, we use ground-truth data from PeeringDB, the largest source of self-reported data about the properties of ASes. From PeeringDB, we extract the self-reported business type of each AS, which is from "Cable/DSL/ISP", "NSP", "Content", "Education/Research", "Enterprise", and "Non-profit". We combine the "Cable/DSL/ISP" and "NSP" classes into a single class "Transit/Access". We ignore the "Non-profit" category for the purposes of this classification. The labeled ground-truth data thus consist of three classes (Table 1): "Transit/Access", "Content", and "Enterprise". As PeeringDB under-represents the "Enterprise" category, we manually assemble a set of 500 networks which we determine to be enterprise customers based on their WHOIS records and webpages, and add this set to the labeled classification data. The dataset contains the business-type associated with each AS. Our AS classification dataset was obtained from CAIDA on 04/01/2021 and contains 71,665 ASes.

File format: $\langle AS \rangle \mid \langle Source \rangle \mid \langle Class \rangle$

Table 1:   AS classification ground-truth dataset.

Source	Description
CAIDA-class	Classification was an inference from the machine-learning classifier
peerDB-class	AS classification was obtained directly from the PeeringDB database

Class	Description
Transit/access	ASes which were inferred to be either transit and/or access providers
Content	ASes which provide content hosting and distribution systems
Enterprise	Various organizations, universities, and companies at the network edge that are mostly users, rather than providers of internet access, transit, or content

3.4.   AS relationship experimental dataset

We obtained BGP paths from Route Views and RIPE NCC on the first day of April in 2014–2018 as our source data. After unified pre-processing, three series of link relationship inference data containing labels were obtained using three existing inference techniques. We used the intersection of the result sets obtained by the three inference methods on the same day as the experimental data for subsequent experiments. The reason is as follows.

(1) The accuracy of each relationship inference approach has generally been more than 90% in recent years, which means that the existing methods have mastered the general features of the internet topology structure. Similar to previous study,[19] only using the inference result of a certain method as a dataset must have a bias.

(2) The scale of the internet is growing, and the number of route collectors and VPs is also increasing. This means that the data in more recent years are better in terms of quantity and structural integrity. The amount of data determines the model's ability to express the real situation to a certain extent.

(3) The intersection of the inference results of multiple technologies can be seen as a method of hard voting based on probability. Our vote focuses on the result consistently. Based on the results of previous researches, we have greatly reduced the error of the dataset we constructed.

4. Features Description and Analysis

4.1. Basic feature

Although most interconnection agreements between networks are secret, some information can be inferred from traceroute-derived and BGP-derived internet topology datasets at IP, router, and AS level. The dataset obtained above enables the measurement of the influence of autonomous systems in the global routing system, the inference of relationships between AS, and other functions. The basic features we use for AS classification are also derived from these datasets and are summarized as follows.

(1) Customer, provider, and peer degrees: We obtain the number of customers, providers, and peers (at the AS-level) using AS relationship experimental dataset (Section 3.4).
(2) Size of customer cone in number of ASes: We obtain the size of an AS' customer cone using AS relationship experimental dataset (Section 3.4).
(3) Size of the IPv4 address space advertised by that AS: we obtain this quantity using BGP routing tables collected from Route-views.[31]

4.2. Features Analysis

Before edge feature analysis, we calculate the differences of features between the target AS and the source AS to represent the corresponding link as follows:

$$\Delta = |f_{AS1} - f_{AS2}|, \qquad (1)$$

where f_{AS1} and f_{AS2} are the feature values of AS1 and AS2, respectively, and Δ is the difference. This method is used by default in the feature analysis that follows.

4.2.1. *Degree and transit degree*

Degree is one of the most basic indices in graph theory to describe the importance of a node. The degree of node i in an undirected network is defined as the number of edges directly connected to it. *Transit degree*[15] is the number of unique neighbors appearing in transit paths of an AS by extracting triplets. For example, assume the existence of paths *AS1, AS2, AS3, AS4* and two triplets *(AS1, AS2, AS3), (AS2, AS3, AS4)*. Since both *AS2* and *AS3* have two different neighbor nodes from the extracted triples, their *transit degrees* are both 2. Besides, ASes with a transit degree of zero are *stub* ASes, which are at the outermost layer of the network (i.e., *AS1* and *AS4*). The *transit degree* is more suitable for describing the relationship cluster of ASes in the internet, which mainly reflects the scale of the customer service and cooperation of an AS. Therefore, *degree* and *transit degree* are used as important graph structure features for subsequent algorithm design.

4.2.2. *Distance to clique*

Inferring clique, which is the AS at the top of the hierarchy and forms a transit-free state with each other (can be seen as P2P relationship), is the first step of extracting the *distance to clique* feature. With the detailed steps of selecting clique ASes, we refer to the previous work.[15] The *distance to clique* feature is mainly to capture the distance from the ASes to the network center. This feature is based on the assumption that high-tier ASes generally have a large number of customers, so they are easier to be peers to achieve mutual benefits. The better strategy for the ASes at the low-tier are to rely on the top ASes to achieve global accessibility and form provider–customer (P2C) relationship finally.

We first construct an undirected graph by extracting the AS links in the BGP paths as edges. After that, we calculate the average shortest path distance from each AS to the clique ($\frac{\sum_{i=1}^{N} D_i}{N}$, where N is the number of the clique and D_i is the shortest distance from an AS to the ith AS in the clique) and map it to the corresponding link using Eq. (1). Finally, we use the 12/01/2020 validation dataset to evaluate the four types of AS links, and the distribution of *distance to clique* is shown in Fig. 3(a). This feature well reflects the significant difference between P2P and P2C relationships. This is one of the reasons that the P2P and P2C relationships are easy to distinguish using existing inference algorithms.

4.2.3. *Assign VP*

Vantage points (VPs, which can be intuitively understood as the first nodes of AS paths) are typically distributed in many different geographic locations, especially at the upper tiers of the internet hierarchy. Meanwhile, the number of VPs is also very limited, compared with the scale of complete internet structure. We analyze the quantity of VPs, which can detect the same AS link. We visualize the discrimination among different types of AS relationships in Fig. 3(b) about this feature. From the figure, we can observe that more than 97% P2P links can be detected by less than 100 VPs (referring to the previous work[18]). More than half P2C are seen by more than 110 VPs. Hence, for the single feature *assign VP*, the two types of AS relationships (i.e., P2C and P2P) tend to be similar. This result once again confirms the excellence of the features selected by the previous algorithm.

4.2.4. *Common neighbor ratio*

This feature is defined as the ratio of common neighbors between ASes to the total number of neighbors, which is a new key point. Intuitively, due to business, the P2C relationship is that, in most cases, an AS as a customer cannot efficiently skip an AS as a merchant to achieve global reachability. Instead, it needs to establish a connection with an AS as a merchant. Therefore, the overlap rate of AS at both ends of a P2C link is very low. On the contrary, for P2P, as a kind of peer cooperation and mutual benefit relationship, it has high similarity between ASes. Therefore, P2P's common neighbor ratio is higher than another type of ASes. As shown in Fig. 1(c), the proportion of P2P relationships with large common neighbor ratios both exceed 60%, which validates our intuitive idea. This new feature once again verifies the high discrimination of the two relationships.

4.2.5. *Distance to VP*

Different from *distance to clique*, we also pay attention to the distance from each node (AS) to the first node (VP) in each BGP path.

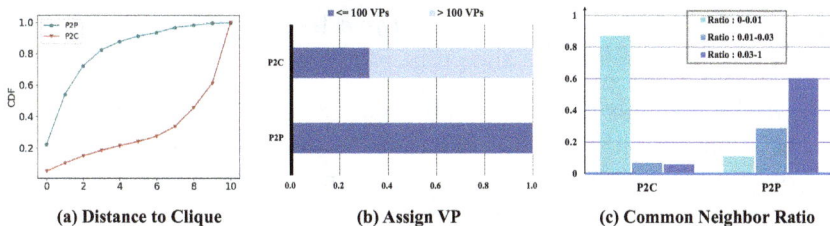

(a) Distance to Clique (b) Assign VP (c) Common Neighbor Ratio

Fig. 1: Analysis of the *Distance to Clique, Assign VP, Common Neighbor Ratio*. (a) CDF of absolute distance between ASes to clique for different relationships. (b) The distribution of the number of VPs with a threshold 110 that can be detected on each relationship's links. (c) The distribution of different common neighbor rates for different relationships.

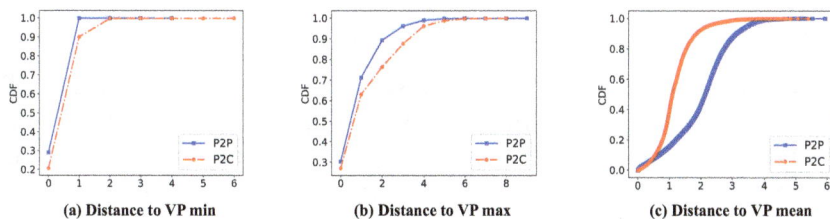

(a) Distance to VP min (b) Distance to VP max (c) Distance to VP mean

Fig. 2: CDF of the *Distance to VP* using different calculation methods on two types of AS relationships. (a) The minimum distance of the edge from the VP in which the edge can be observed. (b) Maximum distance. (c) Mean distance.

This feature indicates that we expect to count the distance set from the target AS to VP in all BGP paths to reflect the position of a link in many paths. Because the same node can appear in several paths, the *distance to VP* value of the node is expressed as a set of integers. In the face of these integer sets, the mean value of the set represents the universality of the node position, and the maximum and minimum values represent the specificity of the node position. As shown in Fig. 2, we can observe that using the mean value of the set is more discriminative among the two types. In the following feature importance analysis (see Fig. 2), it also proves that the importance of the mean value is higher than the maximum and minimum values.

4.2.6. *Node hierarchy*

The internet obeys a certain hierarchical structure.[37] Considering the hierarchical features of ASes, we pay attention to the distribution of different types of AS links in the internet topology. We refer to the work of *K-Shell*[37] and use *transit degree* to decompose the internet into three components (as shown in Fig. 3(a)):

(1) All ASes in the clique form a nucleus, which belongs to the smallest clusters. However, the ASes in the clique have a large average degree, and most AS links started from the clique connect to the outer structure of the internet topology.
(2) The rest of the ASes with zero transit degrees are considered as the shell of the internet. Simultaneously, there are cases where some ASes are directly connected to the core. This part constitutes the largest component.
(3) The remaining ASes are all classified into one category.

4.2.7. AS type

An organization has related business types due to its functions. *AS type* has been considered to be a very important feature, since it has a direct impact on the AS relationship. We get the AS classification dataset from CAIDA.[38] The ground-truth data are extracted from the self-reported business-type of each AS list in PeeringDB.[36] After that, *AS type* can be summarized into three main categories: (1) *Transit/Access*. This type of ASes are inferred to be either a transit and/or access provider. (2) *Content*. This type of ASes provide content hosting and distribution systems. (3) *Enterprise*. This type of ASes include various organizations, universities, and companies. Furthermore, we also add the fourth type: (4) *Unknown*. This group contains those ASes that don't have a clear type, and the neutral

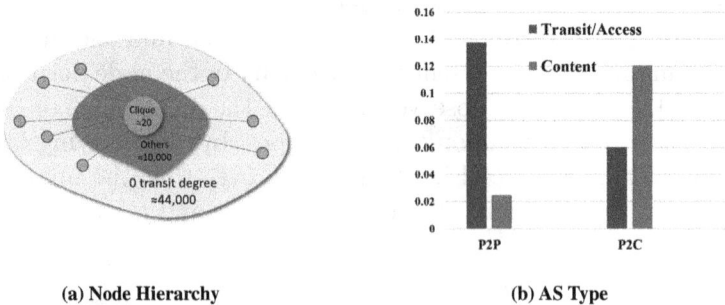

(a) Node Hierarchy　　　　　　　　(b) AS Type

Fig. 3: (a) Three components of the Internet based on *K-Shell*. (b) The distribution of the two AS types between P2P and P2C.

ASes that do not belong to the first three categories. We take the type of the source node of each edge as the feature of the edge, and the results of the two types of edges as shown in Fig. 3(b).

5. Methodology

In this section, based on the idea of aggregating surrounding information, we propose ASGNN, a graph neural network-based framework, to achieve AS classification under complex internet structures. We input graph structure and feature information from the previous section into our model for training and compare it with ground truth to minimize the negative log-likelihood loss. Therefore, the same model can be used to solve the important classification problem in the internet network.

5.1. *Aggregate surrounding information*

In the traditional AS classification algorithms, they generally regard each AS as a separate sample and assign corresponding features. However, the internet, as a typical complex network, has its corresponding meaning for every node and every edge. Therefore, if we only treat each node as an independent sample, we ignore a lot of important feature information. In summary, we conduct AS classification experiment based on AS-level network topology to explore whether the information between nodes has specificity and correlation to the characteristics of nodes themselves. In that way, we could benefit by adopting graph neural network (GNN)[39] to utilize more structural information beyond first-order neighbors for classification.

From the row vector perspective of matrix multiplication, graph-based convolution process is equivalent to the aggregation operation of the feature vectors of the neighbor nodes. The process of graph convolution operation can realize efficient filtering operation on graph data. Moreover, GNN brings powerful fitting capabilities by stacking and modifying multiple GNN layers.

5.2. *Problem description*

For AS classification problem, we are given a graph $G = (V, E, W)$ with a subset of nodes $V_l \subset V$ labeled, where V is the set of n nodes in the graph (possibly augmented with other features), and $V_u = V \backslash V_l$ is the set of unlabeled nodes. Here W is the weight matrix, and E is the set of edges. Let y be the set of m possible labels, and $Y_l = \{y_1, y_2, ..., y_l\}$ be the initial labels on nodes in the set V_l. The task is to infer labels Y on all nodes V of the graph.

5.3.   Model framework

ASGNN mainly consists of six types of layers, i.e., input layer, feature layer, GNN layer, MLP layer, optimization layer, and output layer.

Input layer: The source dataset of the input has been elaborated in Section 4. Faced with different tasks, the first step is to build an undirected and weighted graph $G = (V, E, W)$, which represents a static internet AS-level network, where V denote the set of ASes, $E \in V \times V$ denotes the set of business relationships between ASes, and W corresponds to the weight of the set of edges E. Let $v_i \in V$ denote an AS and $e_{ij} = (v_i, v_j) \in E$ denotes an AS link pointing from v_i to v_j. The adjacency matrix $\mathbf{A}$ is an $n \times n$ matrix with $\mathbf{A}_{ij} = 1$ if $e_{ij} \in E$ and $\mathbf{A}_{ij} = 0$ if $e_{ij} \notin E$. The graph has node attributes $\mathbf{X}$, where $\mathbf{X} \in \mathbb{R}^{n \times d}$ denotes the feature matrix and $x_v \in \mathbb{R}^d$ represents the feature vector of node v.

Feature layer: As illustrated in Fig. 4, the feature layer is mainly to construct its own feature vector for each node and generate a corresponding weight value for each edge. For AS classification task, the feature vector is composed of two parts: AS classification features and AS relationship features. AS classification features are the

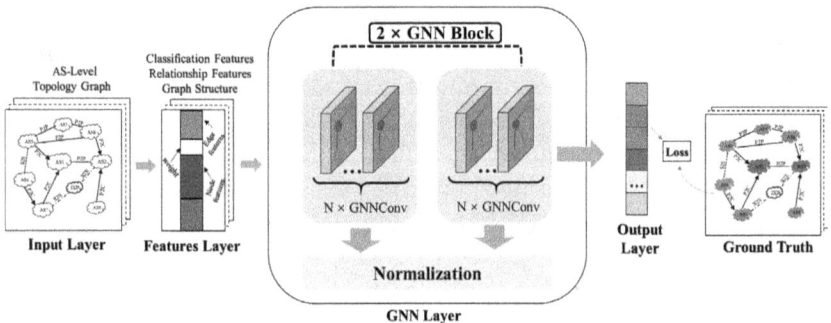

Fig. 4:   Model framework of ASGNN.

three types presented in Section 4.1. AS relationship features are the seven important features mentioned above (i.e., *degree, transit degree, distance to clique, distance to VP, assign VP,* and *node hierarchy*). *Common neighbor ratio* constitutes the weight of the edges.

GNN Layer: In the previous section, we have introduced the basic idea of using graph convolution operation. GNN is a semi-supervised learning algorithm for graph structured data, which makes use of Laplace transform to make the node aggregate the features of higher-order neighbors. In particular, GNN instantiates $\mathbf{A}$ to be a static matrix closely related to the normalized graph Laplaican to get the following expression:

$$\hat{\mathbf{A}} = \tilde{\mathbf{D}}^{-1/2} \tilde{\mathbf{A}} \tilde{\mathbf{D}}^{-1/2}, \qquad (2)$$

where $\tilde{\mathbf{A}} = \mathbf{A} + \mathbf{I}$, $\tilde{\mathbf{D}}_{ii} = \sum_j \tilde{\mathbf{A}}_{ij}$, with $\mathbf{A}$ being the adjacency matrix and $\mathbf{I}$ being the identity matrix. $\hat{\mathbf{A}}$ can be regarded as a graph displacement operator.

GNN carries out convolution operation in the spectral domain, and each operation can aggregate an additional layer of features. Spectral convolution function is formulated as

$$\mathbf{H}^{(l+1)} = \sigma\left(\hat{\mathbf{A}} \tilde{\mathbf{H}}^{(l)} \mathbf{W}^{(l)} \right), \qquad (3)$$

where $\mathbf{W}^{(l)}$ denotes the layer-specific trainable weight matrix and $\sigma(\cdot)$ is a nonlinear activation function (i.e., Sigmoid function). Moreover, $\mathbf{H}^{(l)} \in \mathbb{R}^{n \times k}$ represents the matrix of activation at l layer, where n and k denote the number of nodes and output dimensions of layer l, respectively. In particular, we set $\mathbf{H}^{(0)} = \mathbf{X}$ as initialization input. Each convolution operation would capture the neighbor's features of additional layer. If the objects of the first matrix multiplication are $\mathbf{A}$ and $\mathbf{X}$, then they are equivalent to the nodes combined first-order neighbor features. The more such multiplications, the more layers of information that are abstractly merged.

We design a model containing two identical blocks, each with two-layer GNN, to achieve semi-supervised node classification on a graph with a symmetric adjacency matrix $\mathbf{A}$. Our forward model then takes the block's simple form

$$f(\mathbf{X}, \mathbf{A}) = \text{Norm}\left(\text{ReLU}\left(\hat{\mathbf{A}} \, \text{ReLU}\left(\hat{\mathbf{A}}\mathbf{X}\mathbf{W}^{(0)}\right)\mathbf{W}^{(1)}\right)\right), \quad (4)$$

where $\mathbf{W}^{(0)}$ is an input-to-hidden weight matrix and $\mathbf{W}^{(1)}$ is a hidden-to-output weight matrix. ReLU is the activation function and normalizes the result. The neural network weights $\mathbf{W}^{(0)}$ and $\mathbf{W}^{(1)}$ are trained using gradient descent.

Optimization layer: Let $\mathbf{Z}$ denote the output of GNN layer. For AS classification task, we apply the cross-entropy loss as the objective function. Formula (5) is applied to AS classification:

$$\mathcal{L}(\theta) = -\sum_{v_i \in V} z_i \log\left(Z_i\right), \quad (5)$$

where z_i denotes the label of node v_i, and θ denotes all parameters needed to be learned in the model.

Output layer: The model output is compared with ground truth to minimize the negative log-likelihood loss. In the course of the experiment, the training set is used to train the model to determine the weight and bias of the model, the validation set makes the model in the best state by adjusting the hyperparameters, the testing set used only once during the entire experiment is used to evaluate the performance of our model. We save the model with the best experimental result.

6.   Evaluation

In this section, we evaluate our GNN-based inference algorithm, ASGNN, on the experimental dataset. We have compared the performance with many machine learning methods such as support vector machines, decision tree, Xgboost, LightGBM, etc. We have proved the performance of ASGNN in the following three aspects:

(1) Our proposed GNN-based model achieves an accuracy of 74% in the AS business-type classification task, which significant by outperforms all comparative methods.
(2) The features based on AS relationships have a significant enhancement on AS type classification.

(3) The experimental comparison shows that using our proposed common neighbor ratio values as the edge weights of the AS graph can further enhance the classification performance of the ASGNN model.

(4) Through the parameter optimization and feature importance analysis of the model, the performance of the model can be explained to a certain extent.

6.1. *Baseline methods*

In order to evaluate the effectiveness of the model, our model is compared with the eight different methods. These baselines are described as follows.

(1) **Support vector machines (SVM)** is a linear classifier to maximum margin on the feature space.

(2) **Naive Bayes (NB)** is a classification method based on Bayes's theorem and the assumption of conditional independence of features.

(3) **KNN** is a supervised classification algorithm implemented by distance.

(4) **Logistic regression (LR)**, as a classical classification model, is popular in industry for its simplicity, parallelizability, and strong interpretability.

(5) **Decision tree (DT)** is a conditional probability distribution defined on the feature and class space.

(6) **MLP** is a three-layer feed-forward model using back propagation.

(7) **Xgboost**[40] is one of the Boosting algorithms. The idea of the Boosting algorithm is to integrate many weak classifiers together to form a strong classifier.

(8) **LightGBM**[41] is a new Gradient Boosting Decision Tree (GBDT) implementation with two novel techniques: Gradient-based One-Side Sampling (GOSS) and Exclusive Feature Bundling (EFB).

6.2. *Evaluation metrics*

In order to evaluate the performance of the proposed model, the following metrics are used: *accuracy, precision, recall,* and *F1*. Because the sample is imbalanced, it is difficult to reflect the performance of the model if the *Accuracy* is used as the only evaluation metric. Thus, we also consider *precision*, which is the number of the samples predicted to be positive are truly positive samples, and *recall*, which indicates how many positive examples in the original sample were predicted correctly. *F1* combines the cases of *precision* and *recall* to evaluate the performance of the classification model. The mathematical derivation of these parameters is illustrated using the following equations:

$$\text{Accuracy} = \frac{\text{TP} + \text{TN}}{\text{TP} + \text{TN} + \text{FP} + \text{FN}}, \tag{6}$$

$$\text{Precision} = \frac{\text{TP}}{\text{TP} + \text{FP}}, \tag{7}$$

$$\text{Recall} = \frac{\text{TP}}{\text{TP} + \text{FN}}, \tag{8}$$

$$F1 = 2 \times \frac{\text{Precision} \times \text{Recall}}{\text{Precision} + \text{Recall}}, \tag{9}$$

where TP, TN, FP, and FN refer to *true positive, true negative, false positive,* and *false negative,* respectively.

Our model is implemented using PyTorch.[42] The parameters are updated by Adam algorithm.[43] Each experiment runs 1,000 epochs in total. In the AS-type classification, all results are obtained by training the ASGNN using Adam with weight decay 5×10^{-4} and an initial learning rate of 0.1. We use two blocks, where each block has two standard GNN layers (i.e., ASGNN setting can be represented as 2×2), to learn the graph structure.

In summary, we select the best parameter configuration based on performance on the validation set and evaluate the configuration on the testing set.

6.3.   *AS classification result*

The initial AS-type inference model only considered two or three features and used traditional machine learning algorithms. The current algorithms have several times more relevant features. Based on the features mentioned in Section 4, we set up four groups of experiments. Each set of experiments compares the results of our proposed ASGNN model with other models to reflect our performance. The details of these four situations are as follows:

(1) **Situation 1**: 11 regular AS classification features, and the weight value of all edges of the AS graph is 1.
(2) **Situation 2**: 11 regular AS classification features and 8 AS relationship-based features, the weight value of all edges of the AS graph is 1.
(3) **Situation 3**: 11 regular AS classification features, and the weight of the edge is determined by feature *Common Neighbor Ratio*.
(4) **Situation 4**: 11 regular AS classification features and 8 AS relationship-based features, the weight of the edge is determined by feature *Common Neighbor Ratio*.

AS classification results are shown in Table 2. ASGNN behaves surprising well in inferring the three kinds of types, with accuracy, precision, recall, and $F1$ is generally at the best (The bolded numbers in the table). Focusing on $F1$ metrics, from NB, LR, DT, Xgboost, MLP, and ASGNN(W) methods, we found that adding relevant features of AS relationship to AS type classification task

Table 2:   The *Accuracy*, *Precision*, *Recall*, and *F1* of the eight feature-based methods and our ASGNN model for AS classification experiment.

Models	11 Features				19 Features			
	Acc.	Pre.	Rec.	$F1$	Acc.	Pre.	Rec.	$F1$
SVM	65.0	42.0	65.0	51.0	65.0	42.0	65.0	51.0
NB	30.0	68.0	30.0	17.0	31.0	69.0	31.0	18.0
KNN	65.0	67.0	65.0	64.0	59.0	59.0	59.0	58.0
LR	65.0	49.0	65.0	51.0	65.0	63.0	65.0	59.0
DT	62.0	53.0	62.0	51.0	64.0	54.0	64.0	53.0
LightGBM	63.0	57.0	63.0	52.0	63.0	57.0	63.0	52.0
Xgboost	62.0	50.0	62.0	51.0	70.0	68.0	70.0	68.0
MLP	59.0	67.0	59.0	58.0	62.0	65.0	62.0	61.0
ASGNN	66.0	**90.0**	66.0	**75.0**	67.0	82.0	67.0	72.0
ASGNN(W)	**72.0**	79.0	**72.0**	**75.0**	**74.0**	**85.0**	**74.0**	**78.0**

Note: The bold entries represent the best value of the current metric among the many methods.

can improve AS classification performance to a different extent by comparing experiments. The results of this experiment corroborate our speculation that the type of business of AS and the relationship established between ASes are mutually influential. ASGNN (W) differs from ASGNN in that the former AS graph is an undirected weighted graph with weights derived from the common neighborhood ratio feature. With the addition of weights, the metrics of ASGNN(W) are significantly better than ASGNN. With the ASGNN model considering weights and with 19 features as input, the $F1$ metric reaches 78%, which is a substantial improvement compared to all comparison methods.

In summary, a series of experiments demonstrate the superiority of combining GNN into the AS-type classification task and the positive correlation effect between AS classification and AS relationships.

6.4. *Feature importance analysis*

In this subsection, we divided the 19 features into two categories, AS classification features and AS relationship features, and performed a feature importance analysis using the decision tree algorithm.

Figure 5 shows the bar chart of importance results for the 19 features. It is obvious that the *anno_numberAddresses* in AS classification features play an important role in the classification task. More noteworthy is that the importance of several AS relationship features

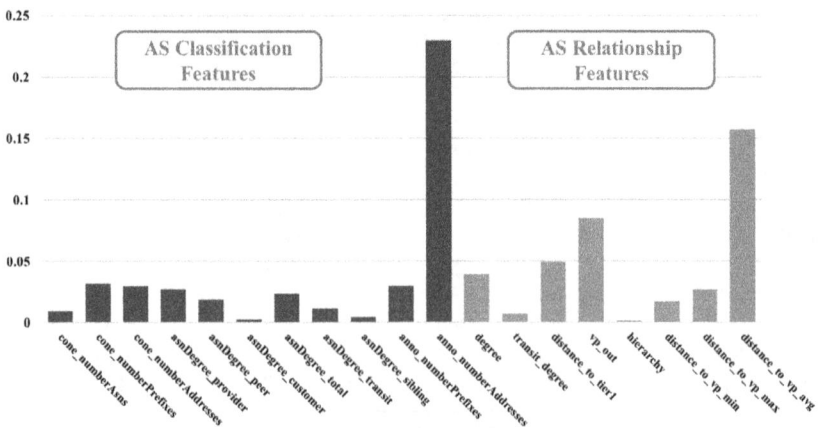

Fig. 5:　Feature importance analysis.

we introduced is significantly better than most traditional AS classification features, such as *distance_to_vp_avg* and *vp_out*, etc. The results of this experiment once again verify our hypothesis that there is a strong correlation between AS classification and AS relationship.

7. Conclusion and Future Work

With the in-depth study of previous inference algorithms and the importance of relevant AS features, we find that there is a close relationship between AS classification and AS relationship. Thus, through experiments, we verify that the results of using AS relationship in AS classification task can have positive effects. Meanwhile, we propose a GNN-based model, ASGNN, which achieves outstanding results compared with many baselines on the highly reliable datasets we construct.

In the future, on the question of the AS relationship, we will focus more on the inference of complex relationships (hybrid relationships and partial transit relationships) with machine learning methods. In terms of AS classification, we believe that sample imbalance is always an important reason for the poor prediction performance of model classification, followed by the lack of more critical features. This will also be the direction of our subsequent experiments.

References

1. P. Gill, M. Arlitt, Z. Li, and A. Mahanti, The flattening internet topology: Natural evolution, unsightly barnacles or contrived collapse? in *International Conference on Passive and Active Network Measurement*, 2008, pp. 1–10.
2. C. Labovitz, S. Iekel-Johnson, D. McPherson, J. Oberheide, and F. Jahanian, Internet inter-domain traffic, *ACM SIGCOMM Computer Communication Review*. **40**(4), 75–86 (2010).
3. S. Sundaresan, X. Deng, Y. Feng, D. Lee, and A. Dhamdhere, Challenges in inferring internet congestion using throughput measurements, in *Proceedings of the 2017 Internet Measurement Conference*, 2017, pp. 43–56.
4. A. Dhamdhere, D. D. Clark, A. Gamero-Garrido, M. Luckie, R. K. Mok, G. Akiwate, K. Gogia, V. Bajpai, A. C. Snoeren, and K. Claffy, Inferring persistent interdomain congestion, in *Proceedings of the 2018*

Conference of the ACM Special Interest Group on Data Communication, 2018, pp. 1–15.

5. J. M. Smith and M. Schuchard,Routing around congestion: Defeating DDOS attacks and adverse network conditions via reactive BGP routing, 2018, in *2018 IEEE Symposium on Security and Privacy (SP)*, 2018, pp. 599–617.

6. S. Cho, R. Fontugne, K. Cho, A. Dainotti, and P. Gill, BGP hijacking classification, in *2019 Network Traffic Measurement and Analysis Conference (TMA)*, 2019, pp. 25–32.

7. A. Cohen, Y. Gilad, A. Herzberg, and M. Schapira, Jumpstarting BGP security with path-end validation, in *Proceedings of the 2016 ACM SIGCOMM Conference*, 2016, pp. 342–355.

8. P. Gill, M. Schapira, and S. Goldberg, Let the market drive deployment: A strategy for transitioning to bgp security, *ACM SIGCOMM Computer Communication Review*. **41**(4), 14–25 (2011).

9. J. Karlin, S. Forrest, and J. Rexford, Autonomous security for autonomous systems, *Computer Networks*. **52**(15), 2908–2923 (2008).

10. M. Apostolaki, G. Marti, J. Müller, and L. Vanbever, Sabre: Protecting bitcoin against routing attacks, arXiv preprint arXiv:1808.06254 (2018).

11. M. E. Tozal, Policy-preferred paths in as-level internet topology graphs, *Theory and Applications of Graphs*. **5**(1), 3 (2018).

12. L. Gao, On inferring autonomous system relationships in the internet, *IEEE/ACM Transactions on Networking*. **9**(6), 733–745 (2001).

13. X. Dimitropoulos, D. Krioukov, M. Fomenkov, B. Huffaker, Y. Hyun, K. Claffy, and G. Riley, As relationships: Inference and validation, *ACM SIGCOMM Computer Communication Review*. **37**(1), 29–40 (2007).

14. E. Gregori, A. Improta, L. Lenzini, L. Rossi, and L. Sani, BGP and inter-as economic relationships, in *International Conference on Research in Networking*, 2011, pp. 54–67.

15. M. Luckie, B. Huffaker, A. Dhamdhere, V. Giotsas, and K. Claffy, As relationships, customer cones, and validation, in *Proceedings of the 2013 Conference on Internet Measurement Conference*, 2013, pp. 243–256.

16. J. S. Varghese and L. Ruan, A machine learning approach to edge type prediction in internet AS graphs. *Comput. Sci. Techn. Rep*. **375**, 9 (2015).

17. Y. Jin, C. Scott, A. Dhamdhere, V. Giotsas, A. Krishnamurthy, and S. Shenker, Stable and practical {AS} relationship inference with problink, in *16th {USENIX} Symposium on Networked Systems Design and Implementation ({NSDI} 19)*, 2019, pp. 581–598.

18. Z. Jin, X. Shi, Y. Yang, X. Yin, Z. Wang, and J. Wu, Toposcope: Recover as relationships from fragmentary observations, in *Proceedings of the ACM Internet Measurement Conference*, 2020, pp. 266–280.

19. T. Shapira and Y. Shavitt, Unveiling the type of relationship between autonomous systems using deep learning, in *NOMS 2020-2020 IEEE/IFIP Network Operations and Management Symposium*, 2020, pp. 1–6.

20. V. Giotsas, M. Luckie, B. Huffaker, and K. Claffy, Inferring complex as relationships, in *Proceedings of the 2014 Conference on Internet Measurement Conference*, 2014, pp. 23–30.

21. R. Govindan and A. Reddy, An analysis of internet inter-domain topology and route stability, in *Proceedings of INFOCOM'97*, 1997, vol. 2, pp. 850–857.

22. Z. Ge, D. R. Figueiredo, S. Jaiswal, and L. Gao, Hierarchical structure of the logical internet graph, in *Scalability and Traffic Control in IP Networks*, 2001, vol. 4526, pp. 208–222.

23. L. Subramanian, S. Agarwal, J. Rexford, and R. H. Katz, Characterizing the internet hierarchy from multiple vantage points, in *Proceedings. Twenty-First Annual Joint Conference of the IEEE Computer and Communications Societies*, 2002, vol. 2, pp. 618–627.

24. X. Dimitropoulos, D. Krioukov, G. Riley, *et al.*, Revealing the autonomous system taxonomy: The machine learning approach, arXiv preprint cs/0604015 (2006).

25. A. Dhamdhere and C. Dovrolis, Twelve years in the evolution of the internet ecosystem, *IEEE/ACM Transactions on Networking.* **19**(5), 1420–1433 (2011).

26. R. Motamedi, R. Rejaie, and W. Willinger, A survey of techniques for internet topology discovery, *IEEE Communications Surveys & Tutorials.* **17**(2), 1044–1065 (2014).

27. S. Y. Qiu, P. D. McDaniel, and F. Monrose, Toward valley-free inter-domain routing, in *2007 IEEE International Conference on Communications*, 2007, pp. 2009–2016.

28. G. Di Battista, M. Patrignani, and M. Pizzonia, Computing the types of the relationships between autonomous systems, in *IEEE INFOCOM 2003. Twenty-Second Annual Joint Conference of the IEEE Computer and Communications Societies* (IEEE Cat. No. 03CH37428), 2003, vol. 1, pp. 156–165.

29. CAIDA, Center for Applied Internet Data Analysis, https://www.caida.org/.

30. J. Friedman, T. Hastie, R. Tibshirani, *et al.*, Additive logistic regression: A statistical view of boosting (with discussion and a rejoinder by the authors), *Annals of Statistics.* **28**(2), 337–407 (2000).

31. University of Oregon, Route views project, http://www.routeviews.org/.

32. R. NCC, Routing information service (RIS), https://www.ripe.net/ analyse/internet-measurements/, created: 04 Mar 2015.
33. IANA, Internet Assigned Numbers Authority, https://www.iana.org/ assignments/as-numbers/as-numbers.xhtml.
34. E. Katz-Bassett, D. R. Choffnes, Í. Cunha, C. Scott, T. Anderson, and A. Krishnamurthy, Machiavellian routing: Improving internet availability with BGP poisoning, in *Proceedings of the 10th ACM Workshop on Hot Topics in Networks*, 2011, pp. 1–6.
35. CAIDA, Inferred as to organization mapping dataset, 2011, https:// www.caida.org/catalog/datasets/as-organizations/, published August 3, 2014.
36. PeeringDB, https://www.peeringdb.com/.
37. S. Carmi, S. Havlin, S. Kirkpatrick, Y. Shavitt, and E. Shir, A model of internet topology using k-shell decomposition, *Proceedings of the National Academy of Sciences*. **104**(27), 11150–11154 (2007).
38. CAIDA, As classification dataset, https://www.caida.org/catalog/ datasets/as-classification/, published September 10, 2015.
39. T. N. Kipf and M. Welling, Semi-supervised classification with graph convolutional networks, arXiv preprint arXiv:1609.02907 (2016).
40. T. Chen and C. Guestrin, Xgboost: A scalable tree boosting system, in *Proceedings of the 22nd ACM SIGKDD International Conference on Knowledge Discovery and Data Mining*, 2016, pp. 785–794.
41. G. Ke, Q. Meng, T. Finley, T. Wang, W. Chen, W. Ma, Q. Ye, and T.-Y. Liu, Lightgbm: A highly efficient gradient boosting decision tree, *Advances in Neural Information Processing Systems*. **30**, 3146–3154 (2017).
42. A. Paszke, S. Gross, F. Massa, A. Lerer, J. Bradbury, G. Chanan, T. Killeen, Z. Lin, N. Gimelshein, L. Antiga, and A. Desmaison, Pytorch: An imperative style, high-performance deep learning library. *Advances in Neural Information Processing Systems*, 2019.
43. D. P. Kingma and J. Ba, Adam: A method for stochastic optimization, arXivpreprint arXiv:1412.6980 (2014).

https://doi.org/10.1142/9789811266911_0010

Chapter 10

Social Media Opinions Analysis

Zihan Li and Jian Zhang[*]

Institute of Cyberspace Security, Zhejiang University of Technology,
Hangzhou, P.R. China

[*]*zhangjian-hdu@hdu.edu.cn*

With the rapid growth of the internet in the last decade, online social media has become one of the major channels for information spreading. Individuals can exchange their opinions on various news on platforms such as Chinese Toutiao. However, such free-flowing information could also provide grounds for violent behaviors. Most existing studies ignore the interaction among comments and the corresponding replies. In this chapter, we propose an end-to-end model PathMerge. This model is applied for controversy detection. PathMerge method takes both dynamic and static features into consideration, and then integrates the information from the graph structure of the social network dynamic information. Experiments on the real-world dataset demonstrate that our model outperforms existing methods. Analysis of the results prove our model has significant generalization ability.

1. Introduction

Social media such as Chinese Toutiao[a] have become major channels where people share their views. In the open and free circumstance, exchanging different opinions might lead to fierce discussions and

---

[a]https://www.toutiao.com.

even war of words. This pollutes the cyber environment. The cause of these controversies can be political debates[1,2] or other topics.[3] The contents of such comments represent certain public sentiment. It provides opportunities to solve the major problems in network governance, such as news topic selection, influence assessment, and polarized views alleviation.[4]

Therefore, controversy detection in social media has drawn attention.[2,5] Existing methods for controversy detection in social media focus mainly on macro-topic controversy detection. For example, some of the specific topics in Twitter can raise large-scale controversial debates among different users.[6,7] Moreover, for the data collected from news portals, researchers pay much attention to whether certain news items are likely to raise conflicts among users.[8] The existing methods mainly detect and analyze conflicts from a macro perspective. We concentrate on detecting the conflicts between the comments under the news with a certain topic.

In this chapter, we detect the micro controversy among comments from social media. According to recent research,[9] controversial comments always have debatable content and express an idea or an opinion which generates argument in the response. This represents an opposing opinion in disagreement with the current comment. Figure 1 gives an example of a controversy over a certain piece of news. The news N belongs to topic T and it is followed by multiple

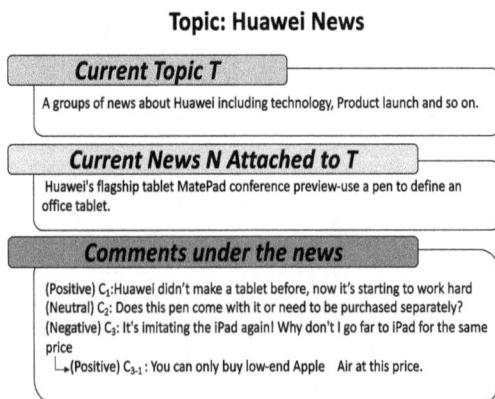

Topic: Huawei News

Current Topic T
A groups of news about Huawei including technology, Product launch and so on.

Current News N Attached to T
Huawei's flagship tablet MatePad conference preview-use a pen to define an office tablet.

Comments under the news
(Positive) C_1:Huawei didn't make a tablet before, now it's starting to work hard
(Neutral) C_2: Does this pen come with it or need to be purchased separately?
(Negative) C_3: It's imitating the iPad again! Why don't I go far to iPad for the same price
$\hookrightarrow$(Positive) C_{3-1} : You can only buy low-end Apple   Air at this price.

Fig. 1:   A piece of news under the topic of Huawei, and the comments under the news. Each comment is labeled as positive, negative, or neutral, depending on its attitude to the news.

comments which show different opinions. The comments are labeled as positive, neutral, or negative based on their attitudes to the current news. There exists a situation that C_{3-1} expresses refutation literally while it actually supports N. This is because in the comment tree, it refutes C_3, a refuting comment to N. So we mark comment C_{3-1} as a positive comment.

The general approach to measure controversy is context dependent and uses the texts to make controversial detection.[10] It develops a controversy detection pipeline and the method is evaluated on the data collected from Twitter. Moreover, Sriteja *et al.*[11] consider more specific information based on social media, such as like and share numbers of one comment on the Facebook platform. It integrates these features to predict people's reactions to a piece of growing news or issue. Instead of using information based on the social reaction, Zhao *et al.*[12] exploit pattern expressions to discover rumors in tweets for rumor detection. Although methods using natural language process (NLP) can predict rumors or controversial content directly, some phenomena can easily cause semantic recognition errors. The sensations coming from the same content can be diverse. For example, *That is interesting* in English. When a person shows interest in a topic, he can use *That is interesting* to express his positive attitude toward the current topic. However, when he is not interested in a topic, *That is interesting* can also express his perfunctory attitude. The only way to distinguish its real meaning is from the context and the tone of speakers. Therefore, controversy detection based on semantic analysis suffers from such ambiguous texts.

Recently, graph embedding methods have achieved great success in many areas due to their ability to encode complex networks from high dimensions into embedding vectors in lower dimensions.[13] It can also encode features of nodes from complex graphs.[14,15] We propose a model named PathMerge (see Fig. 2) which integrates structure features with dynamic features of the heterogeneous graph including nodes of news, comments, and users. First, as Fig. 3 shows, we create a graph structure to describe the relationship among comments, users, and news. To preserve the reply-structure information of the comment tree, we connect each comment node with its parent comment node, and the news node with the first-level comment node. To include the information from user to comment, we connect each comment node with its poster user node. Then, a random

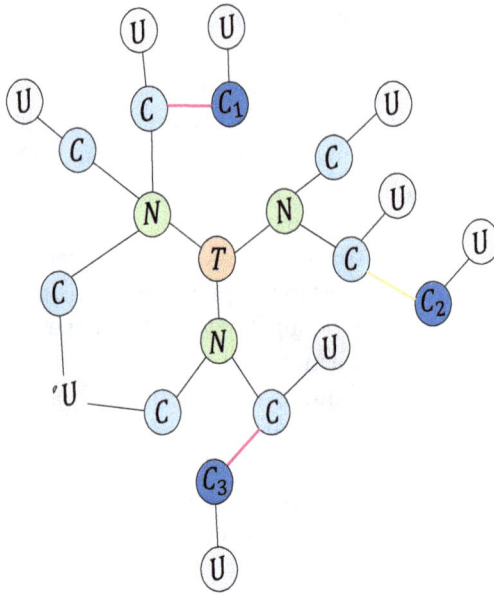

Fig. 2: The architecture of graph construction. T refers to the topic of the news. N represents one news under the topic. C and U refer to the comment and the user. The red line from comment node to comment node means controversy while the blue line means non-controversy.

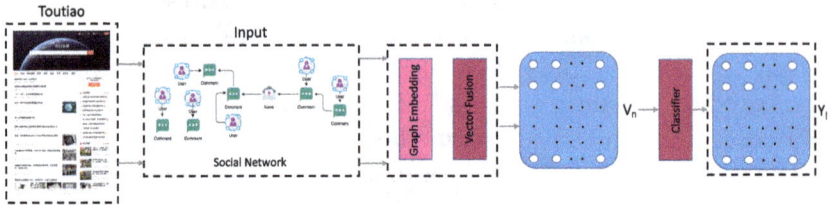

Fig. 3: The architecture of PathMerge method framework, V_n represents the embedding vectors of every node, and Y_l stand for the prediction result.

walk-based model is applied to derive the node representation of the reply-structure information. Finally, the embedding vectors of aimed comments and their replied comments are merged to predict the existence of controversy between comments.

PathMerge method mainly uses global graph structures in the process of obtaining feature vectors. We propose a novel feature mixing method in the process of feature calculation, which integrates the feature vectors of nodes among the specific path in the comment tree. The specific path indicates the route from the root node to the leaf node. We obtain the final feature vectors of edges through three different formulations combining several node feature vectors. This process aims to fuse the feature information of the nodes in the specific path. Experimental results demonstrate that our model outperforms baselines. The main contributions of this chapter are as follows:

1. We build a Chinese dataset for controversy detection consisting of 511 news items, 71.579 users, and 103,787 comments under the same topic collected from Chinese Toutiao, each of the comments is labeled as controversial or non-controversial.
2. We propose a random walk-based model, PathMerge, for comment-level detection. The model can integrate the information from the nodes of the path from the root node to the current node in the aimed comment tree. Especially, PathMerge can further fuse the dynamic features.
3. Extensive experiments on Toutiao dataset demonstrate the temporal and structural information can effectively improve the embedding vectors and get a better result in AUC and AP metrics. Also, our model performs generalizability under different ratios of training samples.

The rest of this chapter is organized as follows. In Section 2, we review previous studies on controversy detection and several graph embedding methods. In Section 4, we describe the PathMerge method in detail and explain the construction of such heterogeneous graphs. In Section 5, we evaluate the AUC and AP metrics using PathMerge against other baselines. In Section 6, we conclude this chapter and highlight some future research directions.

2. Related Work

Controversy detection has been studied on web pages or social media for many years.[16-18] Existing methods on detecting controversy in media mostly aim at detecting controversial articles in blogs or other web pages. Early methods on controversy detection are mainly based on statistical features, such as revision times[19] and user edit history,[20–22] etc. Others incorporate the sentiment-based features[20] and semantic features.[23] Existing web-based works usually exploit the controversy on Wikipedia[17,24] and user comments[25] for detection.

Unlike web pages, social media consist of more diverse topics and more fierce discussions among users, which makes controversy detection on social media more challenging. Early studies assume that a topic has its inner controversy under related topics, and focus on topic-level controversy detection. An early experiment[6] to detect controversial snapshots (consisting of many tweets referring to a topic) was based on Twitter. Graph-based method[2] builds graphs based on a Twitter topic, such as retweeting graphs and following graphs. Then it applies graph partitioning to measure the extent of the controversy. However, comment-level detection is rough, because there are many wrongly spelled words and polysemy cases. Recent works focus on post-level controversy detection by leveraging language features, such as emotional and topic-related phrases,[26] emphatic features, and Twitter specific features.[7] Other graph-based methods exploit the features from the following graph and comment tree.[1] The limitations of current post-level works is that they do not effectively integrate the information from the content and reply structure and ignore the role of posts in the same topic. Moreover, the difference between intra-topic and inter-topic modes is not realized. Hesse *et al.*[4] deal with the topic transfer. They train on each topic and test on others to explore the transferability, which is not feasible in practice.

2.1. *Graph embedding*

Graph embedding techniques have been proved effective in network analysis recently. They map nodes, links, or graphs to low-dimensional dense representations, making it possible to apply machine learning methods for tasks such as node classification,[27] link prediction,[28] graph classification,[29] etc.

DeepWalk[30] samples node sequences through random walk. It incorporates skip gram which was first introduced in *natural language processing* (NLP) to learn node embedding. It first samples a amount of node sequences by applying a walking strategy starting at different nodes and then embeds each node into a vector space. To capture both local and global structural features, node2vec[13] extends

DeepWalk by employing breadth-first search (BFS) and depth-first search (DFS) when sampling node sequences. The sampling strategy is called biased random walk, which indeed improves the ability of network representation learning. Node2vec holds hyper-parameter p, and q influences the probability of node visiting. DeepWalk and node2vec are two methods developed for Homogeneous network. Other random walk-based graph embedding methods may have similar ideas but adopt different sampling strategies which focus on different aspects of the network structure.[31,32] The emergence of deep learning methods accelerated the growth of this typical research area. Among the variants of graph neural networks the (GNN), GCN[33] provided a simplified method to compute the spectral graph convolution to capture the information of graph structure and node features, and transform the structural information between nodes into vectors. GAE[15] designs a decoder to restructure relational information of the nodes. The decoder uses the embedding vectors obtained by GCN to reconstruct the graph adjacency matrix. Then it performs iteration according to the loss of the reconstructed adjacency matrix and label matrix. In practice, GAE achieves better link prediction results than other algorithms.

All the works discussed above focus on a static network where the nodes and edges do not change over time. With the development of graph embedding, temporal graphs have been proposed by many researchers. A large amount of works have explored the temporal phenomena in graphs: StreamWalk,[34] which is based on the concept of temporal walks, updates the node embedding online to track and measure node properties and similarity from a graph stream. CTDNE[35] learns directly from the temporal network (graph stream) without having to approximate the edge stream as a sequence of discrete static snapshot graphs. Tempnode2vec[36] generates PPMI matrices from individual network snapshots to optimize a loss function and capture temporal stability.

3. Toutiao Dataset

In this section, we first give a detailed description of the dataset and data preprocessing. Then we conduct preliminary descriptive analyses.

3.1. *Data description*

Our main source of data is the Toutiao social platform. The collected data cover multiple aspects under the theme of Huawei (technology, entertainment, finance, etc.). We collect the news from March 2019 to December 2019 related to the Huawei topic and the corresponding comments content and user name under the news. We manually evaluate the comments to check whether they are positive, negative, or neutral through multiple annotators. To classify them, we consider the content of the comments and the responses they received. If the content of a comment is controversial and expresses an idea or a point of view, this comment is marked as controversial. Moreover, some of the comments are mainly for publishing advertisements, we directly mark these comments as neutral. We label comments with three sentiments: positive, neutral, and negative. Each comment is labeled by at least two annotators. When a disagreement occurs, the third annotator joins. All of the labels given by the annotators are considered and determined by the fourth and fifth people. Due to the large amount of the original dataset, we divide the original dataset into three subsets. However, there are some abnormal data in the dataset. For example we note some special comments that neither have content nor belong to any user. For this situation, we assume this comment is deleted by the author. As Fig. 4 shows, one

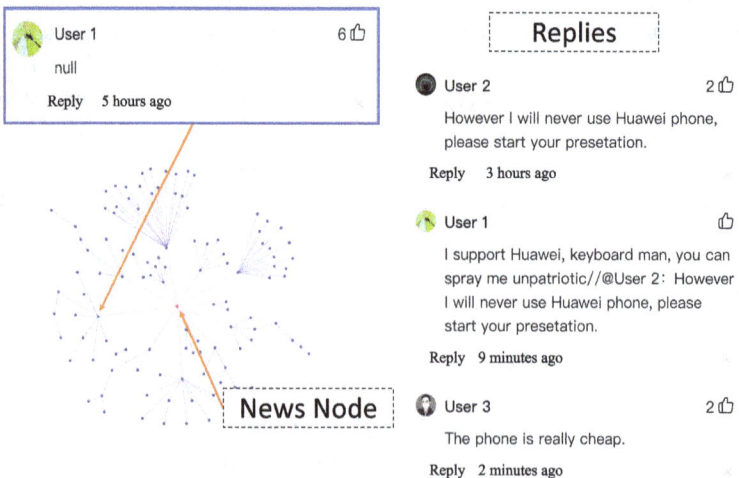

Fig. 4:   A controversial comment tree under one news of Huawei, with one of the comments having been deleted.

of the first-level comments has been deleted and its content becomes unknown. In this situation, we can only conclude its label through its child-comments. For Fig. 4 case, most of the child-comments under the comment are against Huawei and do not follow the comments' view. In that case, we think this comment presents supportive attitude to a current Huawei topic.

3.2. Label

We use Huawei-related news in Toutiao social media and bring in a third-party organization to mark the data of the comments. The principles are as follows:

(1) The marked data can only have five values at most, which are marked by five different people.
(2) Five individuals label the comment in order. If two people give the same opinion on a comment consecutively, the labeling of this comment is finished. It is considered that these kinds of comments are clear, and do not require all five people to judge on then.
(3) If the comments given by two consecutive people are labeled with different marks, the labeling process continues. The comment tendency obtained by the above process is finally calculated by the following formula to get its comment tendency:

$$result = \frac{1}{n} \sum_{i=0}^{n} s_i, \tag{1}$$

$$L = \begin{cases} -1 & \text{if } r < -0.3 \\ 0 & \text{if } r < 0.3 \quad \text{and} \quad r > -0.3, \\ 1 & \text{if } r > 0.3 \end{cases} \tag{2}$$

s represent a set of all the scores given for the comment. And s_i means the i-th item of this set. n is the length of this score-set. L means the last result of the classifier. As for L, -1 represents a negative comment, indicating a negative opinion on Huawei under the current topic, 0 represents a neutral comment, indicating an unclear opinion on Huawei under the current topic, and 1 represents a positive comment, indicating a supportive attitude toward Huawei under the current topic.

3.3.   Pre-processing

Based on the original dataset, we first extract all news id, comment id, and user ids. Then we encode every node to rebuild a new network. We also calculate the time difference between the posted time of child-comment and the parent-comment nodes. For the top comment node in the comment tree, we use the time difference between the created time of the news and the top comment posted time. Besides, there are also some comments which contain no information about the text, the posted time, and the posted user. For such comments, we have to find the child comment and the parent comment and use this time difference to infer these comments' posted time. We perform extensive experiments on three subsets of the real-world dataset. Table 1 shows the statistics of the Toutiao dataset. The details are as follows:

Toutiao dataset: We built a Chinese dataset for controversy detection from the Toutiao website for this work. In total, this dataset contains 511 news items, 103,787 comments, and 71,579 users. After data preprocessing, there are 55,994 comments, 494 news items, and 28,908 users left from the total datasets.

Given such a large amount of data, we would like to focus on the controversies under hot news in this chapter. Thus, we sample three subsets of the data for experiments. Specifically, we first find the top two active users who posted the most comments under different news items and denote them as $u1$ and $u2$. The news items commented by $u1$ and the corresponding comments consist of one subset, namely Toutiao#1. Another subset, namely Toutiao#2, consists of the news

Table 1:   The statistics of Toutiao dataset.

Item	Amount
News	511
Comments	103,787
Users	71,579
Positive comments	54,994
Negative comments	25,557
Neutral comments	23,236
Controversial comment pairs	22,184
Non-controversial comment pairs	16,339

Table 2:   Three subsets sampled from Toutiao dataset.

Number	Toutiao#1	Toutiao#2	Toutiao#3
Num. news	11	11	1
Num. user	5,940	3,496	1,573
Num. comment	10,580	5,570	2,466
Num. controversy-replies	4,570	2,610	1,166
Num. non-controversy-replies	2,976	1,309	584
Num. replies	9,504	4,995	2,294
Num. edges	19,685	10,418	4,644

commented on by $u2$ and their comments. The hottest news which we consider as the common news commented on by both $u1$ and $u2$ and their comments comes to form the third subset, namely Toutiao#3. The basic statistics of the whole dataset and its extracted subsets are presented in Table 2. Table 2 shows statistics of the subsets we sampled from Toutiao datasets. Figure 5 is the TSNE result of these three datasets.

3.4.   *Dataset analysis*

Do controversial edges show any significant structural patterns? In this section, we adopt the degree and centrality of two endpoint comment nodes to analyze the controversial edges. We also analyze the depth of comments in our real-world dataset

Degree distribution: Figure 6 presents the depth of comments for three different types of comments. Positive comments and negative comments often have deeper levels than neutral comments. Figure 7 denotes the controversial and non-controversial relationship between comments in four datasets: total dataset, Toutiao#1, Toutiao#2, and Toutiao#3. According to the distribution of the comment edge attributes of the four datasets, we can conclude: (1) The depth of comment-trees which are in controversial relation is often deeper than that of those in non-controversial relation. In the case of very shallow depth, the proportion of non-controversy between comments is higher than controversy between comments; (2) If the depth of non-controversial comments is greater than 10, the depth increases but the proportion does not decrease; (3) The depth of conflicting and

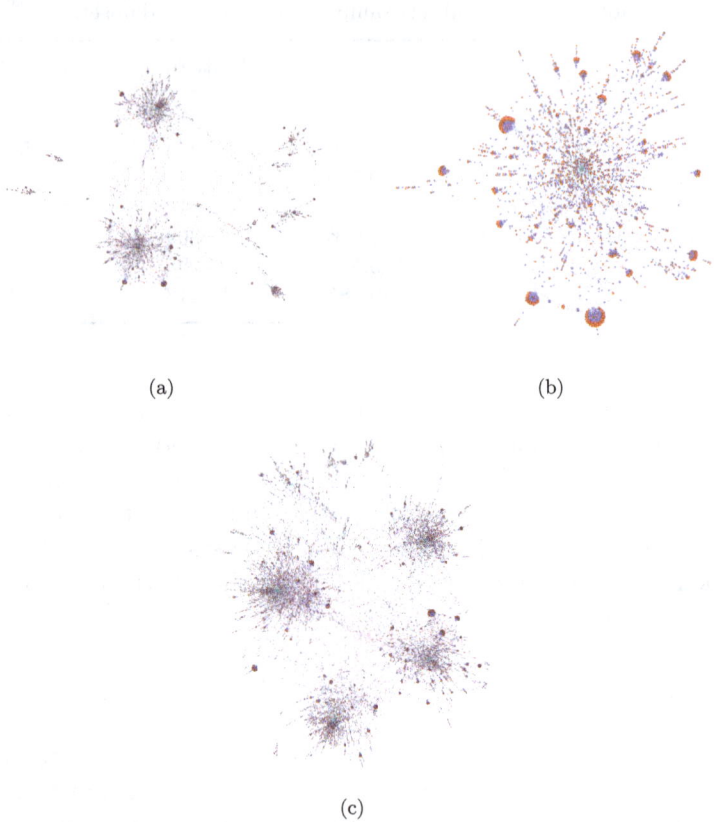

(a)   (b)

(c)

Fig. 5: The three datasets sampled from the total Toutiao datasets. (a) The heterogeneous network of Toutiao#1. (b) The heterogeneous network of Toutiao#3. (c) The heterogeneous network of Toutiao#2.

non-conflicting comment edges in Toutiao#1 is more diverse, so the effect may be better than others; (4) The degree distribution maps of the original network and Toutiao#2, Toutiao#3 are quite different.

Figure 8 presents the distribution of controversial edges with respect to degree and centrality. In Toutiao#1 and Toutiao#2, the majority of controversial edges have larger centrality but lower degree. Part of them have both relatively large centrality and degree. In Toutiao#3, which contains the hottest news, the majority of controversial edges have both large centrality and degree. It implies that there may exist structural patterns of controversial edges, which motivate the graph convolution based solution.

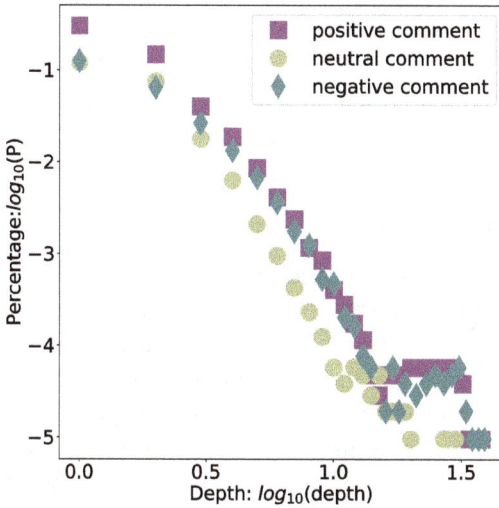

Fig. 6:   The depth of comments in Toutiao comment tree dataset.

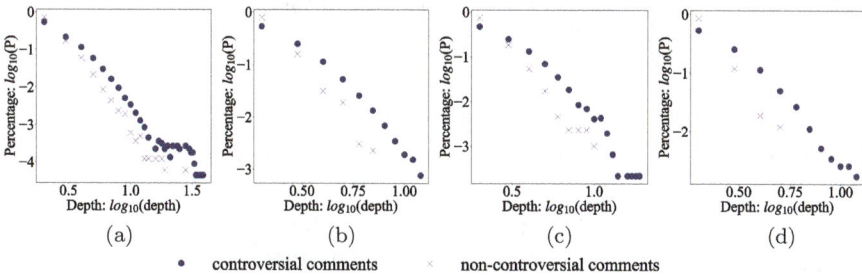

Fig. 7:   The distribution of controversial and non-controversial edges in different depths of the comment tree. (a–d) represent the results in the total dataset, Toutiao#1, Toutiao#2 and Toutiao#3.

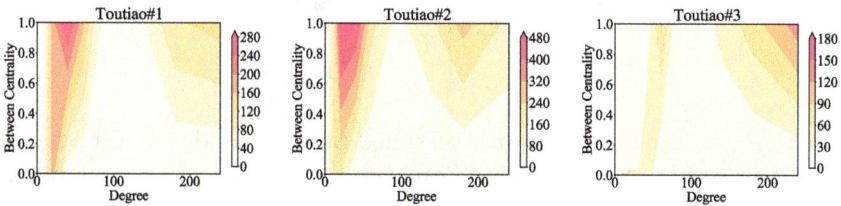

Fig. 8:   Illustration of heatmap distribution modulation consisting of two variables: Node degrees and edge between centrality.

4. Method

In this section, we first introduce the controversy detection problem. Then we show how to apply the proposed PathMerge model to solve the problem in detail.

4.1. *Walking strategy*

In this section, we give a detailed description of the proposed model.

Graph embedding methods have been proved effective for the neural network method. It operates on the graph to encode local graph structures. Node2vec is a widely used graph embedding method. The purposes of node2vec are consistent with our goal. In node2vec, given a source node $\mathbf{u}$, $\mathbf{u}$ can simulate a random walk of fixed length l. Let c_i denote the ith node in the walk, and starting with u, nodes are generated by the following distribution:

$$P(c_i) = x|c_{i-1} = v = \begin{cases} \dfrac{\pi_{vx}}{Z} & \text{if } (v, x) \in E \\ 0 & \text{otherwise} \end{cases}, \qquad (3)$$

where Z is the specified normalization constant.

Node2vec also has two hyperparameters, p and q, to control the random walk strategy. Assuming that the current random walk passes through the edge (t, v) to reach the vertex v, set $\pi_{vx} = \alpha_{p,q}(t, x) \times w_{vx}$, w_{vx} is the weight between vertex v and x:

$$\alpha_{pq}(t, x) = \begin{cases} \dfrac{T_d}{p} & \text{if } d_{tx} = 0 \\ T_d & \text{if } d_{tx} = 1 \\ \dfrac{T_d}{q} & \text{if } d_{tx} = 2 \end{cases}, \qquad (4)$$

where d_{tx} is the shortest path distance between nodes t and x, and T is the result of calculation from time difference. The parameter p indicates the probability of returning to the last walk point and $d_{tx} = 0$ indicates that the vertex t is the vertex x that has just been visited. If the value of P rises, the probability of repeated visits decreases. The parameter q allows the search to differentiate between "inward" and "outward" nodes. When $q > 1$, the random walk is biased to visit the nodes close to the node t (BFS). If $q < 1$, the random walk is biased to visit the node far away from the node t (DFS).

In our model, we introduce another parameter T_d which stands for the dynamic features. The formula is as follows:

$$T_d = \frac{20}{\log_{10}(t_2 - t_1 + 10)}. \tag{5}$$

According to the above walking strategy, the network is traversed. A series of walked node sequences are obtained and vectorized by the Word2vec model.[37] The purpose of this algorithm is to transform text information into vector space. Through this transformation, the connections between words that cannot be directly calculated are transferred to the computable vector space for feature calculation. The purpose is to map the response relationship between nodes that cannot be intuitively quantified to a quantifiable dimensional space representation.

4.2. *Feature calculation*

After obtaining the embedding vector, we develop a new feature calculation technique for integrating vectors.

$$V_r = \alpha_1 \times V_{\text{current}} + \alpha_2 \times V_{\text{topNodes}} + \alpha_3 \times V_{\text{treeNodes}}, \tag{6}$$

where V_r represents the vector calculation result between the current two adjacent comment nodes, V_{current} shows the direct calculation result of the feature vector of the current two adjacent review nodes, V_{topNodes} represents the calculation result of the vector between the child review node in the current two review nodes and the first review node of the entire review tree, α_1, α_2, and α_3 represent 3 constants. As Fig. 9 indicates, if we want to calculate the embedding vector of edge $E_{2,4}$, V_{topNodes} represents the result calculated between comment node 1 and comment node 4. $V_{\text{treeNodes}}$ stands for the total node set from the root node of this comment tree to all nodes passed

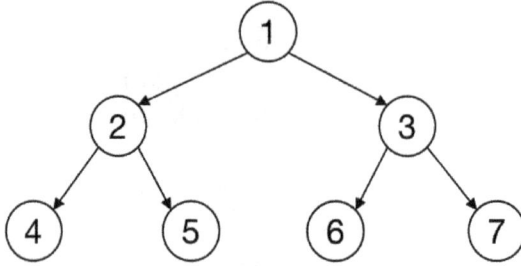

Fig. 9:   The edges of controversy and non-controversy comments' in total dataset, Toutiao#1, Toutiao#2, and Toutiao#3.

by the current node. As for $E_{2,4}$, the total node set is $N = 1, 2, 4$. And the formula of $V_{\text{treeNodes}}$ is

$$V_{\text{treeNodes}} = \frac{1}{m} \sum_{i=1}^{m} (X_{N_i} \times X_{N_{i+1}}) \times 10, \tag{7}$$

where m is the length of the node set N.

5.   Experiment

In this section, we demonstrate the effectiveness of our framework through comprehensive experiments. We present the results on different sub-datasets and the interaction effects through experimental results. We also demonstrate the robustness of our method.

5.1.   Baseline

To validate the effectiveness of our methods, we implement several representative methods including node2vec, metapath2vec, and CTDNE. The basic settings are described as follows:

- **Node2vec**[13] keeps the neighborhood of vertices to learn the vertex representation among networks, and it achieves a balance between homophily and structural equivalence. The ranges of its hyperparameters in this chapter are set to $p, q \in \{0.5, 1, 2\}$.
- **CTDNE**[35] is a general framework for incorporating temporal information into network embedding methods, and is based on random walk, stipulating that the timestamp of the next edge in the walk must be larger than that of the current edge.

- **Metapath2vec**[31] is used for heterogeneous graph embedding based on meta-path walk to get a series of vertex containing various labels, then uses heterogeneous Skip-Gram to generate the embedding vectors. The meta-path strategy is set previously, and it's always symmetrical.

The embedding dimension of node2vec and CTDNE is set to 128. The optimal key parameters p and q of node2vec are obtained through grid search over $\{0.5, 1, 2\}$. For metapath2vec method, the process to classify nodes is as follows: First, we divide all nodes into three types: news, comments, and users, and then we subdivide comment nodes; second, we formulate five meta-paths: including *comments* → *news* → *comments, comments* → *comments, comments* → *users* → *comments, comments* → *comments* → *news* → *comments*, and *comments* → *comments* → *users* → *comments*. Based on these meta-paths, we use the metapath2vec method to map the original network to a 128-dimensional vector space for quantization. For all embedding-based methods, we adopt Logistic Regression (LR) and Random Forest (RF) as the classifier for controversy detection.

5.2. *Performance Comparison*

We compare the performance of PathMerge method with the baselines on the dataset, the evaluation metrics include the macro average precision (Avg. P), macro average recall (Avg. Recall), macro average F1 score (Avg. F1), and accuracy (Avg. Acc). Table 3 shows the performance of all comparing methods and two different classifiers. Node2vec+Time Feature means node2vec method with time

Table 3: The results of total Toutiao dataset.

| Method | Toutiao dataset | | | |
	Avg. Acc	Avg. Recall	Avg. F1	Acg. AUC
Metapath2vec	0.715	0.7217	0.717	0.7152
Node2vec	0.7105	0.7059	0.7129	0.71
Node2vec+Time Feature	0.725	0.7185	0.7291	0.7252
PathMerge	0.7425	0.7437	0.7419	0.7425

features. This is aimed at integrating dynamic features with static structure features during the process of graph embedding.

In the first experiment, we can conclude:

(1) PathMerge outperforms all comparing methods on the total dataset. This indicates that our model can effectively detect controversy with significant generalizability on different datasets.
(2) The structure-based method, node2vec, and metapath2vec are comparable to the other baselines, which proves that structure is necessary for improving the performance.
(3) Dynamic analysis method is helpful to mend the prediction result of controversy in a social network.

In the experiments, our PathMerge method outperforms all baselines by 2.5% of Recall score at least, which validates that it can detect controversy on unseen or dissimilar topics.

And in the next experiment, we divide the total Toutiao dataset into three datasets and compare the performance of PathMerge with the baselines. The evaluation metrics include Area under ROC Curve (Avg. AUC) and the macro average precision (Avg. AP). Table 5 shows the performance of all comparing methods. Note that node2vec and metapath2vec are in two different classifiers. We divide this dataset into the training and testing datasets for the training model and baselines. Table 4 presents the detail of the results among node2vec+Time and PathMerge. We can conclude that the Path-Merge method always achieves the best results.

Table 4: The result of methods' comparison.

Method		Toutiao#1		Toutiao#2		Toutiao#3	
		Avg. AUC	Avg. AP	Avg. AUC	Avg. AP	Avg. AUC	Avg. AP
CTDNE	LG	71.24	68.12	72.51	65.31	74.24	70.46
	RF	81.04	77.19	78.29	73.72	79.50	74.15
Node2vec	LG	71.94	67.36	69.11	65.85	77.21	75.16
(local)	RF	81.14	77.23	77.03	77.22	78.71	74.67
Node2vec	LG	72.23	66.44	71.19	66.26	76.01	76.03
(global)	RF	79.31	74.42	77.75	74.56	78.09	76.23
Metapath2vec	LG	74.06	70.15	71.20	63.87	74.93	71.94
(global)	RF	79.02	75.88	77.82	74.60	77.35	71.52

Table 5: The results of Toutiao#1, Toutiao#2, and Toutiao#3.

Method	Classifier	Avg. Acc	Avg. Recall	Avg. F1	Avg. AUC
		Toutiao Dataset			
Node2vec+Time	Logistic regression	0.613	0.6271	0.5825	0.6049
	Random forest	0.725	0.7185	0.7291	0.7252
PathMerge	Logistic regression	0.651	0.6814	0.6199	0.6554
	Random forest	0.7425	0.7437	0.7419	0.7425

6. Conclusion

In this chapter, we propose the PathMerge method based on graph embedding to integrate the information from the root node to the aimed comment node. Unlike the existing work, we focus on the microcosmic controversy in social networks and exploit the information from related posts on the same topics and the reply structure for more effective detection. We also try to integrate global graph embedding vectors with local graph embedding results. In the Toutiao#2 dataset, PathMerge achieves a better result than both node2vec and PathMerge methods. This suggests that in some news the global graph structure can help to predict local comments' controversy and the controversy mode among them has some common features.

This work can lead to new opportunities for future research. First, the combination of the global network and the graph embedding vector of Toutiao#2 can improve the prediction accuracy. It indicates that there are some specific comment structures in the global network that can assist the prediction of the sub-graph networks. Exploring the characteristics of this structure has certain significance for subsequent prediction of controversy between comments. Moreover, from an application perspective, the number of controversies between comments can be used to generate recommendations and cultivate a healthier "news diet" on social media.

References

1. M. Coletto, K. Garimella, A. Gionis, and C. Lucchese, Automatic controversy detection in social media: A content-independent motif-based approach, *Online Social Networks and Media.* **3**, 22–31 (2017).

2. K. Garimella, G. D. F. Morales, A. Gionis, and M. Mathioudakis, Quantifying controversy on social media, *ACM Transactions on Social Computing.* **1**(1), 1–27 (2018).

3. P. Guerra, W. Meira Jr, C. Cardie, and R. Kleinberg, A measure of polarization on social media networks based on community boundaries, in *Proceedings of the International AAAI Conference on Web and Social Media*, 2013, pp. 8–11.

4. J. Hessel and L. Lee, Something's brewing! Early prediction of controversy-causing posts from discussion features, arXiv preprint arXiv:1904.07372 (2019).

5. H. Lu, J. Caverlee, and W. Niu, Biaswatch: A lightweight system for discovering and tracking topic-sensitive opinion bias in social media, in *Proceedings of the 24th ACM International on Conference on Information and Knowledge Management*, 2015, pp. 213–222.

6. A.-M. Popescu and M. Pennacchiotti, Detecting controversial events from twitter, in *Proceedings of the 19th ACM International Conference on Information and Knowledge Management*, 2010, pp. 1873–1876.

7. A. Addawood, R. Rezapour, O. Abdar, and J. Diesner, Telling apart tweets associated with controversial versus non-controversial topics, in *Proceedings of the Second Workshop on NLP and Computational Social Science*, 2017, pp. 32–41.

8. Y. Mejova, A. X. Zhang, N. Diakopoulos, and C. Castillo, Controversy and sentiment in online news, arXiv preprint arXiv:1409.8152 (2014).

9. M. Coletto, K. Garimella, A. Gionis, and C. Lucchese, A motif-based approach for identifying controversy, in *Proceedings of the International AAAI Conference on Web and Social Media*, 2017, vol. 11, pp. 15–18.

10. J. M. O. de Zarate, M. D. Giovanni, E. Z. Feuerstein, and M. Brambilla, Measuring controversy in social networks through NLP, in *International Symposium on String Processing and Information Retrieval*, 2020, pp. 194–209, Springer.

11. A. Sriteja, P. Pandey, and V. Pudi, Controversy detection using reactions on social media, in *2017 IEEE International Conference on Data Mining Workshops (ICDMW)*, 2017, pp. 884–889. IEEE.

12. Z. Zhao, P. Resnick, and Q. Mei, Enquiring minds: Early detection of rumors in social media from enquiry posts, in *Proceedings of the 24th International Conference on World Wide Web*, 2015, pp. 1395–1405.

13. A. Grover and J. Leskovec, Node2vec: Scalable feature learning for networks, in *Proceedings of the 22nd ACM SIGKDD International Conference on Knowledge discovery and data mining*, 2016, pp. 855–864.

14. S. Pan, R. Hu, G. Long, J. Jiang, L. Yao, and C. Zhang, Adversarially regularized graph autoencoder for graph embedding, arXiv preprint arXiv:1802.04407 (2018).

15. T. N. Kipf and M. Welling, Variational graph auto-encoders, arXiv preprint arXiv:1611.07308 (2016).
16. M. Conover, J. Ratkiewicz, M. Francisco, B. Gonçalves, F. Menczer, and A. Flammini, Political polarization on Twitter, in *Proceedings of the International AAAI Conference on Web and Social Media*, 2011, vol. 5, pp. 17–21.
17. S. Dori-Hacohen and J. Allan, Detecting controversy on the web, in *Proceedings of the 22nd ACM International Conference on Information & Knowledge Management*, 2013, pp. 1845–1848.
18. Y. Choi, Y. Jung, and S.-H. Myaeng, Identifying controversial issues and their sub-topics in news articles, in *Pacific-Asia Workshop on Intelligence and Security Informatics*, 2010, pp. 140–153, Springer.
19. A. Kittur, B. Suh, B. A. Pendleton, and E. H. Chi, He says, she says: Conflict and coordination in Wikipedia, in *Proceedings of the SIGCHI Conference on Human factors in Computing Systems*, 2007, pp. 453–462.
20. B.-Q. Vuong, E.-P. Lim, A. Sun, M.-T. Le, H. W. Lauw, and K. Chang, On ranking controversies in wikipedia: Models and evaluation, in *Proceedings of the 2008 International Conference on Web search and Data mining*, 2008, pp. 171–182.
21. T. Yasseri, R. Sumi, A. Rung, A. Kornai, and J. Kertész, Dynamics of conflicts in Wikipedia, *PloS One.* **7**(6), e38869 (2012).
22. H. S. Rad and D. Barbosa, Identifying controversial articles in Wikipedia: A comparative study, in *Proceedings of the Eighth Annual International Symposium on Wikis and Open Collaboration*, 2012, pp. 1–10.
23. J. Linmans, B. van de Velde, and E. Kanoulas, Improved and robust controversy detection in general web pages using semantic approaches under large scale conditions, in *Proceedings of the 27th ACM International Conference on Information and Knowledge Management*, 2018, pp. 1647–1650.
24. R. Awadallah, M. Ramanath, and G. Weikum, Harmony and dissonance: Organizing the people's voices on political controversies, in *Proceedings of the Fifth ACM International Conference on Web Search and Data Mining*, 2012, pp. 523–532.
25. M. Tsytsarau, T. Palpanas, and K. Denecke, Scalable discovery of contradictions on the web, in *Proceedings of the 19th international conference on World Wide Web*, 2010, pp. 1195–1196.
26. N. Rethmeier, M. Hübner, and L. Hennig, Learning comment controversy prediction in web discussions using incidentally supervised multi-task CNNS, in *Proceedings of the 9th Workshop on Computational Approaches to Subjectivity, Sentiment and Social Media Analysis*, 2018, pp. 316–321.

27. X. Wang, P. Cui, J. Wang, J. Pei, W. Zhu, and S. Yang, Community preserving network embedding, in *Proceedings of the AAAI Conference on Artificial Intelligence*, 2017, vol. 31, pp. 203–209.
28. M. Zhang and Y. Chen, Link prediction based on graph neural networks, arXiv preprint arXiv:1802.09691, 2018.
29. J. B. Lee, R. Rossi, and X. Kong, Graph classification using structural attention, in *Proceedings of the 24th ACM SIGKDD International Conference on Knowledge Discovery & Data Mining*, 2018, pp. 1666–1674.
30. B. Perozzi, R. Al-Rfou, and S. Skiena, Deepwalk: Online learning of social representations, in *Proceedings of the 20th ACM SIGKDD International Conference on Knowledge Discovery and Data Mining*, 2014, pp. 701–710.
31. Y. Dong, N. V. Chawla, and A. Swami, Metapath2vec: Scalable representation learning for heterogeneous networks, in *Proceedings of the 23rd ACM SIGKDD International Conference on Knowledge Discovery and Data Mining*, 2017, pp. 135–144.
32. J. Wang, P. Huang, H. Zhao, Z. Zhang, B. Zhao, and D. L. Lee, Billion-scale commodity embedding for e-commerce recommendation in Alibaba, in *Proceedings of the 24th ACM SIGKDD International Conference on Knowledge Discovery & Data Mining*, 2018, pp. 839–848.
33. T. N. Kipf and M. Welling, Semi-supervised classification with graph convolutional networks, arXiv preprint arXiv:1609.02907 (2016).
34. P. Rozenshtein and A. Gionis, Temporal pagerank, in *Joint European Conference on Machine Learning and Knowledge Discovery in Databases*, 2016, pp. 674–689, Springer.
35. G. H. Nguyen, J. B. Lee, R. A. Rossi, N. K. Ahmed, E. Koh, and S. Kim, Dynamic network embeddings: From random walks to temporal random walks, in *2018 IEEE International Conference on Big Data (Big Data)*, 2018, pp. 1085–1092. IEEE.
36. M. Haddad, C. Bothorel, P. Lenca, and D. Bedart, Temporalnode2vec: Temporal node embedding in temporal networks, in *International Conference on Complex Networks and Their Applications*, 2019, pp. 891–902, Springer.
37. T. Mikolov, K. Chen, G. Corrado, and J. Dean, Efficient estimation of word representations in vector space, in Yoshua Bengio and Yann LeCun, editors, *1st International Conference on Learning Representations, ICLR 2013, Scottsdale, Arizona, USA, May 2–4, 2013, Workshop Track Proceedings*, 2013, pp. 1–12.

Chapter 11

Ethereum's Ponzi Scheme Detection Work Based on Graph Ideas

Jie Jin, Jiajun Zhou, Wanqi Chen,
Yunxuan Sheng, and Qi Xuan*

*Institute of Cyberspace Security, Zhejiang University of Technology,
University of Nottingham Ningbo China, Hangzhou, P.R. China*

*xuanqi@zjut.edu.cn

In recent years, the research and development of blockchain technology
have been prosperous and rapid, promoting the popularity of crypto
currencies. This has attracted a lot of investment in the field of cryptocurrency, but there are have also been many frauds, among which the
Ponzi scheme is a classic example. At present, most of the research on the
Ethereum Ponzi scheme use transaction features and contract codes to
construct manual features and input them into the classifier for training
and classification. This approach ignores the correlation between smart
contracts and investors, which will lead to the loss of some potential
information. In this chapter, we propose a Ponzi scheme detection model
framework based on a graph neural network, which realizes the detection of the Ponzi scheme to a certain extent by extracting the opcode
characteristics of the smart contract itself, account transaction characteristics and interaction characteristics between accounts. Moreover, we
conduct experiments on real Ethereum transaction datasets. The experimental results show that the account interaction information (i.e., network topology information) can improve the accuracy of Ponzi scheme
detection.

1. Introduction

A Blockchain is a distributed shared ledger and database, which is decentralized, tamper-proof, and traceable. These characteristics not only lay the trust foundation of the blockchain but also further create a reliable transaction mode. In other words, blockchain realizes point-to-point trust without the credit endorsement of a third party, and this trust relationship is reflected in two stages, i.e., trust in the authenticity of historical behaviors represented by data on the chain, and trust in future behaviors constrained by rules and mechanisms.

Blockchain has broad application prospects, among which Ethereum has become one of the most popular blockchain platforms. On Ethereum, we can create decentralized applications (DApps) through Ethereum virtual machine language, that is, smart contract, to complete all possible businesses, such as transfer, storage, create sub cryptocurrencies, etc.

In recent years, the vigorous development of blockchain technology has accelerated the popularity of digital currency and brought a prosperous virtual investment market, reflecting its value in technological innovation and industrial revolution. On the other hand, it also enlarges the security problem of encrypted assets, making cryptocurrency crime a constraint in the market supervision. At present, the supervision in Ethereum platform is weak, and no institution or organization provides credit endorsement for it, resulting in frequent illegal and criminal activities such as money laundering, gambling, and phishing fraud, among which the most typical is the Ponzi scheme. The Ponzi scheme was named after Charles Ponzi, who deceived investors with a stamp speculation scheme in the 1920s. It is defined as an investment fraud that pays existing investors with the funds raised by new investors. Nowadays, criminals combine the Ponzi schemes with booming blockchain technology to create more new forms of fraud. As investors' enthusiasm for cryptocurrencies continues to rise, the value of cryptocurrencies such as bitcoin has reached a new high, further attracting many inexperienced investors who blindly enter the market and enabling loopholes for criminals to carry out illegal activities. According to a report published by a company that specializes in cryptoanalysis, virtual currency, and risk analysis, there are more and more frauds, such as Ponzi scheme, ICO revocation, and fraud ICO, which have caused a loss of at least 725 million so far. Therefore, financial security has become an important issue in the blockchain ecosystem, and it is of great significance to study security technology for public blockchain in application scenarios such as risk assessment and market supervision.

At present, existing Ponzi scheme detection methods mainly focus on constructing manual feature engineering on data of contract codes and transaction flow, and further inputting them into machine learning classifiers for model training and account classification. Such approaches ignore the relevance between smart contracts and investors, failing in capturing the interaction pattern of the Ponzi scheme. In this chapter, we construct a transaction network using the data of contract call and transfer transaction in Ethereum, to maximumly preserve the interaction information between smart contracts and investors. Furthermore, we propose a Ponzi scheme detection model based on graph neural network to effectively identify the Ponzi scheme contract accounts in Ethereum. This framework combines the opcode features of the smart contract itself, account transaction features, and account interaction features, achieving the state-of-the-art performance in Ponzi scheme detection in blockchains. The main contributions of this chapter are summarized as follows:

- We comprehensively collect the existing labeled Ponzi and non-Ponzi contract data and construct the **Blockchain Ponzi Scheme Dataset**, which can facilitate the research of Ponzi scheme detection in blockchains.
- We construct a blockchain transaction network using the data of contract call and transfer transaction in Ethereum, to maximumly preserve the account interaction information.
- We propose a Ponzi scheme detection model based on graph neural network, which effectively integrates multiple features including smart contract code, transaction, and account interaction to detect the Ponzi scheme contract accounts in Ethereum.
- Extensive experiments demonstrate the superiority of our method in detecting Ponzi scheme in blockchains.

2. Ponzi Scheme Definition and Detection

2.1. *Ponzi scheme in traditional finance*

The Ponzi scheme is named after Charles Ponzi, who deceived investors with a stamp speculation scheme in the 1920s. Nowadays, the public treats the Ponzi scheme as an illegal financial fraud, and an official introduction proposed by the US Securities and Exchange Commission Investor Network[a] is as follows: *A Ponzi scheme is an investment fraud that pays existing investors with funds collected from new investors. Ponzi scheme organizers often promise to invest your money and generate high returns with little or no risk. But in many Ponzi schemes, the fraudsters do not invest the money. Instead, they use it to pay those who invested earlier and may keep some for themselves. With little or no legitimate earnings, Ponzi schemes require a constant flow of new money to survive. When it becomes hard to recruit new investors, or when large numbers of existing investors cash out, these schemes tend to collapse.*

Early Ponzi schemes were mostly launched offline to attract investment. With the development of technology, scammers turn their fraud methods to the internet, launching Ponzi schemes with "low risk and high return" through telephone calls, social media and online advertisements. However, even with the use of bogus institutions as credit guarantees, Ponzi schemes like this are not highly credible enough to withstand scrutiny.

2.2. *Ponzi schemes on Ethereum*

Ethereum is currently the second most popular blockchain platform after Bitcoin (BTC), also known as blockchain 2.0.[1,2] Developers can create and publish decentralized applications on the Ethereum platform by using programmable smart contracts which allow traceable and irreversible trusted transactions without a third party.[3] The high open source and usability of smart contracts have accelerated the development of the cryptocurrency market. However, it has also become a hotbed of illegal criminal activities. Fraudsters can

---

[a]https://www.investor.gov/protect-your-investments/fraud/types-fraud/ponzi-scheme.

create and deploy fraud-based smart contract projects on Ethereum to easily implement financial fraud and seek illegitimate interests. The feasibility of crime mainly stems from the following points:

- The anonymity and decentralization of the blockchain protect the creators of Ponzi schemes from platform supervision to a certain extent;
- The codes and investment rules of smart contracts are open and transparent, and can no longer be edited or modified after creation and deployment. These characteristics are used by fraudsters to win the trust of investors;
- Many investors are not capable of identifying and diagnosing contract codes, and they blindly invest without verifying the content of the contract, resulting in no return on investment;
- Transactions on the blockchain are frequent and complex, and it is often difficult for regulators to prevent and track fraud.

In order to purify the investment environment of cryptocurrency platforms and reduce the investment losses, researchers have launched lots of meaningful works to detect Ponzi schemes.

2.3. *Ponzi scheme detection on Ethereum*

The research of Ponzi scheme detection on Ethereum mainly concentrates on manual feature engineering. There are two types of features that can be used: contract features and transaction features. Contract features can be extracted from source code, bytecode, and opcode. Among them, the source code is the logic code of the smart contract written in "solidity" language, which is the most original contract code. Bytecode is the compilation result of the source code running on the Ethereum virtual machine, and it is a string of hexadecimal coded byte array, which is called machine language. An opcode is a part of an instruction or field specified in a computer program to perform an operation, most of which has an assembly instruction set and can be obtained by decompiling the bytecode. Transaction features include various contents of transaction records, such as transaction target, transaction timestamp, transaction amount, etc.

Next, we summarize the existing related work according to the different life cycles of the Ponzi contract projects that they act on.

2.3.1. Detect in the initial state

The Ponzi scheme contract project needs to be created first, and the created Ponzi contract only has contract code at the initial time, while there is no contract call and transaction behavior. In this state, all we can use is the contract code. Bartoletti et al.[4] first divided the Ponzi scheme into four types: tree, chain, waterfall, and permission transfer, according to the logic of contract source code. Furthermore, Yu et al.[5] conducted in-depth case analysis on the above four categories, and realized the identification of fraud types based on smart contract source code and transaction characteristics combined with keyword extraction methods. Smart contract code needs to be compiled into bytecode to run on the virtual machine. Bartoletti et al.[4] used Levenshtein Distance to measure the similarity between bytecodes to detect Ponzi schemes. This method relies on known codes to identify similar codes and has certain limitations, since the number of smart contract codes is limited. Lou et al.[6] converted bytecode into the single-channel image and used a convolutional neural network for image recognition to realize Ponzi scheme detection.

2.3.2. Detect in the intermediate state

Once the Ponzi contract project is deployed and started, transaction records are generated gradually. In the intermediate state, the information we can use includes contract code and transaction records. Jung et al.[7] extracted transaction features from transaction records in different time periods, combining with opcode features, and input them to different classifiers for detection. Hu et al.[8] analyzed the investment and return characteristics of smart contracts using transaction records, and trained the LSTM model with contract data for future Ponzi scheme detection. Wang et al.[9] proposed a Ponzi contract detection method based on oversampling LSTM. It considers the contract account features and contract code features at the same time, and uses oversampling strategy to fill in the imbalanced Ponzi contract data.

2.3.3. Detect in the final state

In the final stage of the Ponzi scheme, the contract has a record of all transactions during its life. Chen et al.[10,11] identified financial fraud by analyzing account features and operation code features and using ensemble learning methods such as random forest and XGBoost. Zhang *et al.*[12] used account features, opcode features, and bytecode features, and detected Ponzi scheme contracts via Light-GBM. In another work,[13] they input the same features into the deep neural network model to realize Ponzi scheme detection. Since the Ponzi scheme complies with the Pareto principle, i.e., fraud accounts for only a small part of all acts, there exists an imbalance of positive and negative samples in Ponzi contract data. Some related work has been presented to alleviate the imbalance of label data. Fan *et al.*[14] used the Borderline-smote2 oversampling method to enhance the data in order to expand the Ponzi contract. Wang *et al.*[9] used smote algorithm to improve data label imbalance. Zhang *et al.*[12] used the Smooth-Tomek algorithm for mixed sampling.

3. Feature of Ethereum Data

Due to the fraudulent essence, Ponzi contracts have distinct characteristics compared with normal ones. In this section, we sort out and summarize the characteristics of the Ponzi scheme from three aspects: contract code, transaction flow, and account interaction.

3.1. *Contract code feature*

Ethereum contract code includes the source code,[b] bytecode, and opcode, and their manifestations and related features will be introducted in what folllows. Figure 1 shows the transformation between the three types of code.

- **Source code** is the logical code written in solidity language on Ethereum. Researchers can analyze the functions and execution forms of different contracts based on source code, and obtain key code information of related functions as source code features.
- **Bytecode** is the result of source code compiled by a compiler, and is a string of hexadecimal number-encoded byte arrays, called

---

[b] https://cn.etherscan.com/address/0x311f71389e3de68f7b2097ad02c6ad7b2dde4c71#code.

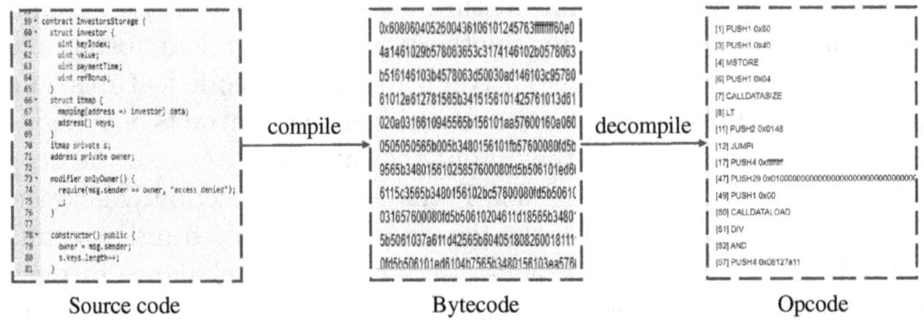

Fig. 1:   Three types of smart contract codes and the transformation between
them.

machine language. Some researchers measure the feature similarity
between the bytecode, which is used to distinguish the Ponzi and
non-Ponzi contracts.

- **Opcode** is the low-level human-readable instructions of the program. All opcodes have their hexadecimal corresponding items,
  e.g., "MStore" and "Sstore" are "0x52" and "0x55", respectively.
  Compared with bytecode, opcode has a certain readability.

In order to verify whether opcode features can play a certain role
in Ponzi scheme detection, we analyzed the feature similarity between
Ponzi contract and non-Ponzi contract. We crawled the bytecode of
the contract from the Etherscan browser and decompiled the bytecode to obtain the opcode. At the same time, considering that EVM
is stack-based, partial opcodes appear in contracts with high frequency. We delete the three most commonly used opcodes, PUSH,
DUP, and SWAP, to reduce their impact on feature analysis. Finally,
we analyzed 76 features of opcodes used for Ponzi contract detection.

We constructed a KNN graph as an opcode feature similarity
network, in which each contract node is connected to its 10 most
similar nodes according to the opcode feature, as shown in Fig. 2.
The red node represents the Ponzi contract, and the blue node represents the non-Ponzi contract. It is obvious that most red nodes
form multiple clusters, indicating that the Ponzi contract opcodes
are diverse. The features in the same cluster also show the commonality of some Ponzi contracts in opcodes. The blue nodes are widely
distributed without obvious clustering distribution, and are related
to some Ponzi contracts, indicating that the codes of non-Ponzi

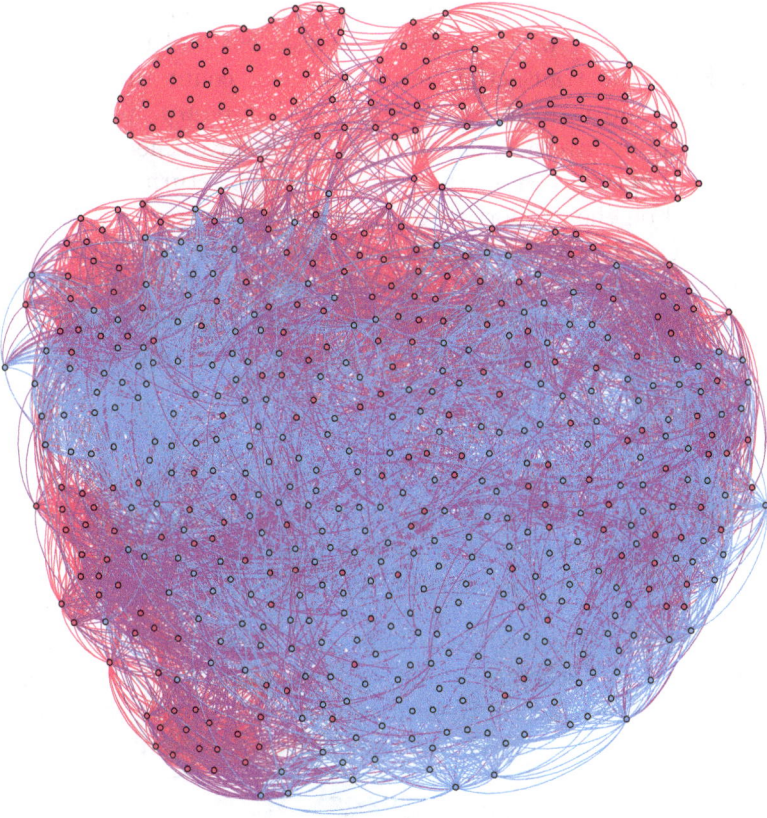

Fig. 2:   Ponzi and non-Ponzi opcode features similarity network.

contracts are diverse, and some contracts are similar to Ponzi ones. In summary, there is a certain degree of distinction between Ponzi and non-Ponzi contracts in opcode characteristics.

3.2.   *Transaction feature*

Since Ponzi scheme is a scam involving investment and returns, we can mine the behavioral information of the Ponzi contract from the transaction flow data. We manually investigate the transaction history of the target smart contracts, and get all transaction records, from which we can directly obtain two types of information: contract transaction amount and contract lifetime.

Particularly, we use the transaction flow direction to measure the transaction-in/out, which represent investment and return, respectively. From the transaction record, we can count the number of investments and returns, i.e, N_invest and N_return, as well as the total, maximum, average, and variance of the investment and return amount (V_invest, V_return). In addition, the return rate is calculated based on the proportion of investors who received at least one payment. For the contract lifetime, we specifically count the lifetime of contracts and mainly calculate the time interval between the initial contract creation time and the latest contract run time. In summary, we obtain 13 transaction features, which can be used as the basic initial features.

3.2.1. Transaction feature analysis

To further understand the expressive power of the extracted features, we investigate the feature importance based on the Mean Decrease Impurity of Random Forest, as shown in Fig. 3.

- *return_rate* plays a key role in detecting Ponzi schemes. The typical nature of Ponzi schemes is that they use funds from new investors to pay interest and short-term returns to old ones to create an

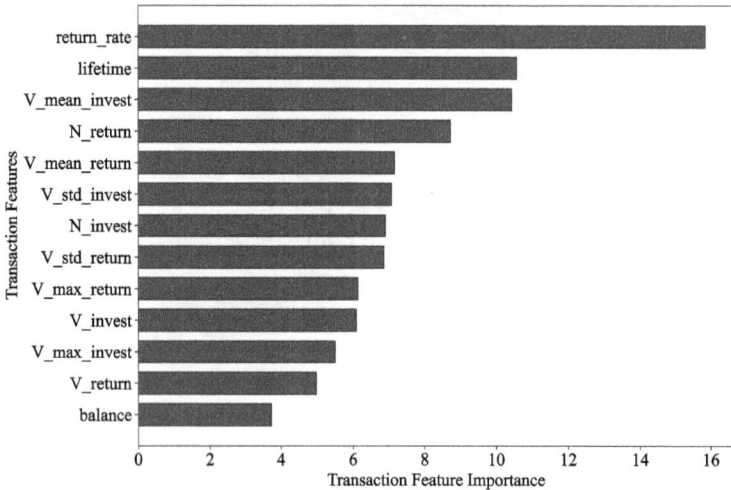

Fig. 3: Importance analysis of manual transaction features via Mean Decrease Impurity of Random Forest.

illusion of profit, which can be observed from the probability that the investor receives the return from contracts, i.e., the return rate.

- *lifetime* of a smart contract includes the process of storage, creation, deployment, execution, upgrade, and destruction. Figure 3 shows that this feature cannot provide a decisive role in detection. This is mainly because only a small number of smart contracts are frequently used and remain active, while most of them have a short lifetime in Ethereum. A Ponzi scheme is often profitable in a short time, it will no longer be used by investors once reported or destroyed in time, resulting in a similar lifetime between Ponzi and non-Ponzi contracts.

- *balance* represents the balance of the contract account during our selected period.

- $V\_mean\_invest/V\_mean\_return$ represents the average investment and return amount in the contract. Ponzi schemes usually set a standard investment amount, while other contracts have relatively no such rigid requirements.

- V_invest/V_return represents the total amount of the investment and the total amount of the return.

- $V\_std\_invest/V\_std\_return$ reflects the concentration or dispersion degree of the investment contract amount and the return amount.

- $V\_max\_invest/V\_max\_return$ represents the largest transaction investment (return) amount in the current contract. Ponzi scheme attracts investors to transfer money to it, which can accumulate a lot of wealth. Meanwhile, the Ponzi scheme will quickly transfer a lot of wealth to another account to avoid scam collapse. Due to the high cost of Ether currency, the transaction amount of most normal contracts is small.

- N_invest/N_return reflects the number of investment or return transactions between two accounts. Ponzi scheme uses the funds of new investors to subsidize the old investors, showing a phenomenon of more input transactions and fewer output transactions. However, with the exposure of the Ponzi scheme, the number of transactions will suddenly decrease or even disappear.

These results suggest that manual features can explain the behavior of smart contracts from different perspectives. Therefore, we extracted these manual transaction characteristics as initial features to facilitate Ponzi scheme detection.

3.3. Network features

The two types of features have been involved in previous studies, and the combination of the two has achieved relatively good performance in the identification of Ponzi schemes. However, the transaction records of blockchains contain rich information and complete traces of financial activities, of which the potential information cannot be fully mined through simple manual features. For example, interaction or behavioral patterns between multiple accounts are difficult to describe through statistical characteristics. Networks are commonly used for describing interacting systems in the real-world, and a considerable part of existing work on cryptocurrency transactions is studied from a network perspective.[15] In this chapter, we collect the transaction records to construct the Ethereum transaction network, where nodes represent the contract accounts and external accounts, and edges represent the existence of a real transaction record between two accounts. Therefore, the problem of identifying the Ethereum Ponzi scheme using transaction network can be treated as a binary node classification problem.

We construct a first-order transaction network of Ponzi and non-Ponzi contracts, and Fig. 4 shows the 5% sampled sub-network. In this network, the red nodes represent the Ponzi contracts, the blue nodes represent the non-Ponzi contracts, and the pink nodes represent the first-order neighbor account of these contracts. Note that these neighbor nodes are usually the investors (i.e., external accounts). As we can see, there are dense connections between these red Ponzi contracts, indicating that investors tend to invest in the same type of contracts.

4. Method

In this section, we first give a detailed description of data collection and preprocessing, and then extract the following meta-features for the smart contracts, which can form more complex features. We also represent the details of our proposed model for Ponzi scheme detection. The framework of our proposed model is illustrated in Fig. 5.

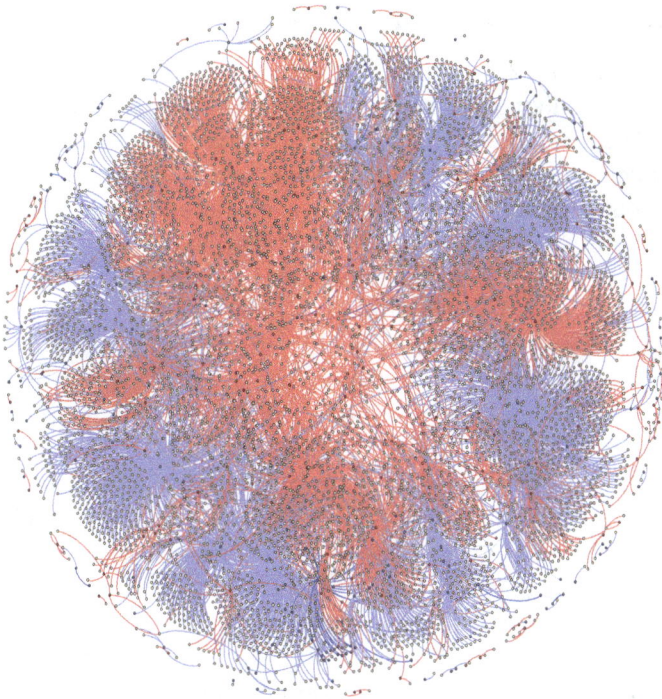

Fig. 4:   A first-order transaction network of Ponzi and non-Ponzi contracts.

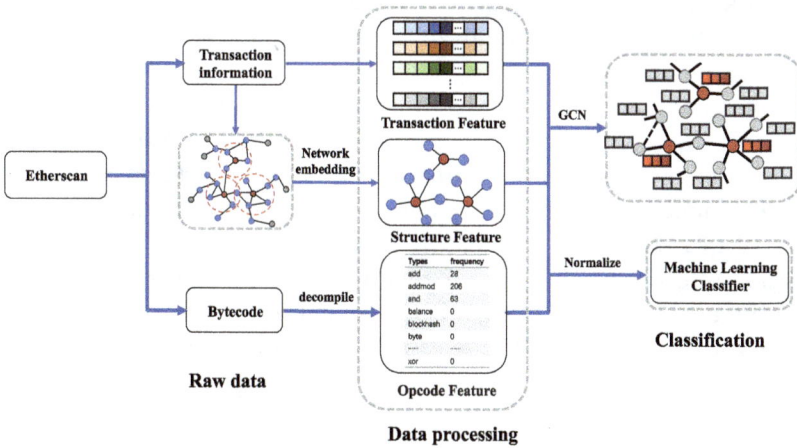

Fig. 5:   Overall framework of our proposed model.

4.1. Data collection and preprocessing

Thanks to the openness of Ethereum, we can easily access transaction records and smart contracts. In this chapter, we first collect 50 labeled Ponzi schemes from an authoritative website named *Etherscan*,[c] which reports various illegal behaviors on Ethereum. However, only 50 contracts labeled as Ponzi schemes are too few to find common characteristics of these smart contracts. We further explore more Ponzi schemes in *Xblock*,[d] which collects datasets for researchers to autonomously access Ethereum meta-datasets. In addition, we also collect 184 Ponzi contracts provided by Bartoletti and others on Google Cloud Disk.[e] We integrate all the above Ponzi contracts, and preprocess these data for building a suitable Ethereum trading network.

We collect transaction records from March 2015 to March 2020. It is worth noting that the transaction records are extremely large, thus we ascertain a number of target smart contracts (Ponzi contracts and normal contracts) and then obtain their transactions from all Ethereum transaction records to make subgraphs for subsequent experiments. As shown in Fig. 6, we randomly sample centered contracts to obtain their 1-hop neighbors and the transaction records between all of them. In other words, we extract the sub-transaction network contained target contracts from the transaction network (the extracted subgraph is highly correlated with these target contracts), and then analyze and identify Ponzi schemes from a network perspective.

For simplicity, the extracted sub-transaction network can be defined as a graph $G = (V, E)$, where $V = \{v_1, \ldots, v_n\}$ refers to a set of accounts (CA and EOA), and $(v_i, v_j) \in E$ represents the transaction among the accounts v_i and v_j. We represent the adjacency matrix, degree matrix, and feature matrix of the sub-transaction network as $A \in \mathbb{R}^{n \times n}$, $D_{ii} = \sum_j A_{ij}$, and $X \in \mathbb{R}^{n \times d}$, respectively, where n is the number of accounts, d is the dimension of the feature vector, and each row $x_i \in \mathbb{R}^d$ is the feature vector of account v_i.

---

[c]https://etherscan.io/.
[d]http://xblock.pro/ethereum/.
[e]goo.gl/CvdxBp.

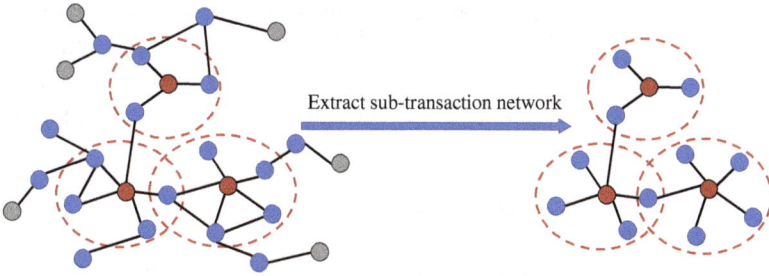

Fig. 6:   Schematic illustration of the sub-transaction network.

4.2. *Model*

In this section, we first introduce the graph convolutional network (GCN) model and then illustrate how to use it for Ponzi scheme detection.

GCN is a semi-supervised convolutional neural network, which can work directly on graphs and use their structural information to help feature learning. The graph convolutional layer is defined as follows:

$$H^{(i+1)} = \sigma \left(\widehat{A} H^{(i)} W^{(i)} \right), \tag{1}$$

where $\widehat{A} = \tilde{D}^{-1/2}(A + I_n)\tilde{D}^{-1/2}$ is a normalized adjacency matrix of the graph with self-connection, $H^{(i)} \in \mathbb{R}^{n \times d_i}$ is the d_i-dimensional representation of n nodes in the ith layer, $W^{(i)} \in \mathbb{R}^{d_i \times d_{i+1}}$ is a weight matrix, $\sigma(\cdot)$ is a nonlinear activation function, and d_{i+1} is the output dimension.

The core of GCN is to collect the characteristic information from all its neighbors for each node, including its own features. However, the above simple GCN model has the following two limitations: (1) In the process of weighted summation, if only the adjacency matrix A is used, the characteristics of the node itself will be ignored. (2) The adjacency matrix A has not been normalized, so multiplying with the feature matrix will change the original distribution of the feature and cause some unpredictable problems. Therefore, we perform a self-loop operation on A, and a standardized treatment of $\tilde{A}$ as follows:

$$\tilde{A} = A + I_n, \tag{2}$$

$$\widehat{A} = \tilde{D}^{-1/2}\tilde{A}\tilde{D}^{-1/2}, \tag{3}$$

where I_n is the identity matrix, A refers to the adjacency matrix, $\tilde{D}_{ii} = \sum_j \tilde{A}_{ij}$, and $\widehat{A} = \tilde{D}^{-1/2}\tilde{A}\tilde{D}^{-1/2}$ is the normalized symmetric adjacency matrix. After that, GCN can capture information about direct neighbors with one layer of convolution. We show the one-layer GCN is defined as follows:

$$H^{(1)} = \sigma\left(\widehat{A}XW^{(0)}\right), \tag{4}$$

where $X \in \mathbb{R}^{n \times d}$ is the initial feature matrix, $W^{(0)} \in \mathbb{R}^{d \times k}$ is a weight matrix, σ is a nonlinear activation function, and $H^{(1)} \in \mathbb{R}^{n \times k}$ is the new k-dimensional node feature matrix, while n and k represent the number of nodes and output size of the layer. Furthermore, we can stack multiple graph convolution layers to extend the receptive field and capture high-order features from graphs. After constructing sub-transaction networks and extracting contract features, we use a two-layer GCN model with softmax function to learn account representations for Ponzi scheme detection:

$$Z = \text{Softmax}\left(\hat{A} \cdot \text{ReLU}\left(\hat{A}XW^{(0)}\right)W^{(1)}\right), \tag{5}$$

where $Z \in \mathbb{R}^{n \times y}$ is the prediction probability distribution and y is the categories of account labels.

5. Experiments

In this section, we first introduce the details of data, comparison methods and experimental settings, and then we present experimental results with discussion.

5.1. *Data*

Some Ponzi contracts only have creation records but no transaction records, which do not meet our requirements. In order to build a connected transaction network that contains as many Ponzi contracts as possible, we screened out contracts for no less than five transaction records. There are 191 Ponzi contracts that meet the requirements. While there are a lot of fraud activities active on Ethereum, only a fraction of them are Ponzi contracts, illustrating an extreme imbalance between the number of Ponzi contracts and normal contracts.

Table 1:   Properties of the sub-transaction networks.

Dataset	$|V|$	$|E|$	K	Proportion	Label rate (%)
Data_ba1	62,151	71,587	2.30		0.61
Data_ba2	48,416	56,175	2.32	191: 191	0.79
Data_ba3	59,109	70,092	2.37		0.64
Data_unba	203,215	304,809	3	191: 1151	0.66

Note: $|V|$ and $|E|$ are the numbers of nodes and edges, respectively, K is the average degree.

In this regard, we constructed two datasets, one of which keeps data balance between Ponzi and non-Ponzi contracts (191:191), and the other keeps the data imbalance (191:1152). And negative sample of balanced dataset are obtained by three random sampling. The details of datasets are shown in Table 1.

5.2.   Comparison methods

Manual feature engineering relies on high-performance downstream classifiers to achieve good task performance, so we use three ensemble learning models, XGBoost Classifier (XGB), Multilayer Perception (MLP), AdaBoost Classifier (ADA), Decision Tree Classifier (DTC), Random Forest Classifier (RFC), Gradient Boosting Decision Tree (GBC), and Logistic Regression (LR), to directly learn the function mapping manual features to account identity.

5.3.   Experimental settings

Regarding the experimental settings, we set up two experimental methods, one is based on non-graph embedding, and the other is based on graph embedding. First of all, the previous scholars who studied Ponzi schemes all studied from the perspective of transaction features and opcode features. So we also use different machine learning classifiers to train in non-graph embedded features and then obtain experimental results. These features do not consider structural information, but only extracted features (transaction feature T, opcode feature O). Secondly, we will discuss whether the proposed graph embedding feature can improve the recognition accuracy of the Ponzi scheme. In order to verify our ideas and introduce the idea of graphs, we use some common embedding methods in graph networks to embed the Ethereum transaction network to obtain a low-dimensional feature vector representation. In particular, the baselines are introduced as follows.

5.4. Evaluation metrics

To measure the effectiveness of different methods in terms of Ponzi scheme detection, we adopt three widely used evaluation metrics, i.e., Precision, Recall, and F1-score, to evaluate our proposed model and other baseline methods.

To ensure a fair comparison, we implement DeepWalk and node2vec using OpenNE[f] (an open-source package for network embedding). For all embedding methods, we set the dimension $d = 16, 32, 64, 128$. For random walk embedding methods, we set hyper-parameters as follows: the size of window $k = 10$, the length of walk $l = 80$, and walks per node $r = 10$. For node2vec, we grid search over $p, q \in \{0.5, 1, 2\}$. For LINE, we use second-order-proximity and set other parameters to the provided defaults. As for our proposed model, we closely follow the framework of Kipf[g] with two-layer GCN, the dimensionality of output is fixed at 128, set the maximum number of epochs to be 500, and setup early stopping. We perform 10-fold cross-validation across all methods and datasets with Random Forest. To evaluate the performance of different methods in terms of Ponzi scheme detection, we consider three evaluation metrics, namely, Precision, Recall, and F1-score. Experiments are repeated for 10 random seed initializations and the average performance is reported.

5.5. Results and discussions

5.5.1. Performance on manual features

We first use manual features to detect Ponzi schemes. Specifically, we first use transaction features (T) or opcode features (O) alone to conduct experiments. Tables 2 and 3 report the detection results of

---

[f]OpenNE: github.com/thunlp/openne.
[g]GCN: https://github.com/tkipf/gcn.

Table 2:   Performance of different classifiers using manual transactionfeatures.

Dataset	Metric	Classifier							
		XGB	SVM	MLP	ADA	DTC	RFC	GBC	LR
Data_ba1	Precision	72.08	70.21	69.89	67.09	72.43	75.32	73.79	69.51
	Recall	75.37	62.72	64.57	76.34	70.48	75.51	78.38	65.70
	F1	**73.11**	65.66	66.65	70.60	70.62	**74.03**	**75.28**	67.05
Data_ba2	Precision	75.70	71.85	70.89	75.14	70.85	79.69	76.25	68.18
	Recall	78.26	63.11	63.87	79.34	76.04	76.83	77.52	68.17
	F1	**76.01**	66.00	66.24	76.07	72.76	**76.48**	**75.93**	66.89
Data_ba3	Precision	72.19	70.03	71.80	70.19	70.58	75.17	72.90	66.97
	Recall	73.19	63.06	64.84	72.28	69.71	74.63	75.90	65.66
	F1	**71.97**	65.80	67.58	70.45	68.97	**74.16**	**73.65**	65.75
Data_unba	Precision	66.56	68.36	70.78	65.37	47.91	75.85	64.91	41.29
	Recall	41.31	32.48	39.13	28.04	48.03	38.08	35.23	10.86
	F1	**50.70**	42.99	**49.39**	38.75	47.63	**50.29**	45.03	16.28

Note: The bold entries represent the best results among the classifiers.

Table 3:   Performance of different classifiers using manual opcode features.

Dataset	Metric	Classifier							
		XGB	SVM	MLP	ADA	DTC	RFC	GBC	LR
Data_ba1	Precision	87.54	75.51	82 47	82.59	84.44	90.73	89.17	77.66
	Recall	88.33	84.54	80 39	85.24	86.82	86.76	85.53	85.12
	F1	**87.57**	79.14	80 65	83.48	84.92	**88.48**	**87.14**	80.32
Data_ba2	Precision	88.93	80.18	83 33	83.31	80.42	90.44	91.51	80.31
	Recall	87.13	86.47	84 42	83.61	83.73	86.73	87.90	86.40
	F1	**87.49**	82.70	83 58	82.84	81.27	**88.34**	**89.26**	82.55
Data_ba3	Precision	88.65	75.97	86 18	86.92	80.12	90.08	89.24	79.35
	Recall	86.17	83.09	84 71	85.01	85.30	83.88	86.75	84.17
	F1	**87.10**	78.57	85 15	85.47	82.28	**86.69**	**87.66**	80.98
Data_unba	Precision	90.34	84.09	89 48	84.94	72.95	94.22	89.83	81.28
	Recall	75.35	69.78	74 93	73.13	74.76	73.70	74.36	66.22
	F1	**81.47**	75.94	**81.01**	78.08	73.12	**82.23**	80.46	72.76

Note: The bold entries represent the best results among the classifiers.

different machine learning classifiers using only transaction features or opcode features, from which we have the following Observations.

Obs. 1. Data balance has a significant impact on the detection based on manual transaction features: Compared with the three balanced datasets, experiments on the imbalanced dataset have a very low recall, indicating that the Ponzi scheme detector based on manual transaction features cannot distinguish Ponzi contracts well in a more realistic scenario. As mentioned earlier, most accounts are normal and their transaction features are diverse, resulting in some normal accounts having similar statistical transaction characteristics to Ponzi ones. Such simple manual transaction features are not capable of capturing the differences in behavior patterns between Ponzi and non-Ponzi contracts.

Obs. 2. Manual features are effective, and opcode features benefit more: Comparing the results in Tables 2 and 3, we can see that a Ponzi scheme detector based on manual opcode features achieves better performance than that with transaction features. In particular, in the case of data imbalance, opcode features still have high detection performance, indicating that there is a significant difference in opcode feature distribution between Ponzi and non-Ponzi contracts. Since the Ponzi contract will definitely meet the characteristics of the scam, it generally implements some single transfer functions, while other contracts have rich functions, such as voting functions, crowdfunding functions, and so on. Therefore, we can effectively identify Ponzi contracts through the opcode.

We then select the top three best classifiers (XGB, RFC, and GBC) for subsequent experiments. Specifically, we further use both transaction features and opcode features to detect Ponzi contracts, and the results are reported in Table 4.

Obs. 3. Random Forest achieves relatively better performance among ensemble learning algorithms across different negative sample selection: As seen in Tables 2 and 3, from the classifiers, only RFC and XGB have a strong generalization and show superior performance on each dataset, while RFC shows the best performance. The reason is that RFC is composed of multiple decision trees, which is an ensemble learning of bagging algorithm, and the classification results are generated by voting. In contrast, XGB and Ada are a kind of ensemble learning of boosting algorithm, and each Ponzi scheme does not have a special correlation relationship, so the effect is poor. Since RFC is a classifier with generalization, it shows better results in different datasets.

Table 4: Performance of different classifiers using transaction and opcode features.

Classifier	T	O	Precision	Recall	F1	Precision	Recall	F1
	Feature		Data_ba1			Data_ba2		
XGB	√		72.08	75.37	73.11	75.70	78.26	76.01
		√	87.54	**88.33**	**87.57**	**88.93**	87.13	**87.49**
	√	√	**87.67**	87.02	86.68	86.69	87.07	86.42
RFC	√		75.32	75.51	74.03	79.69	76.83	76.48
		√	90.73	86.76	88.48	**90.44**	**86.73**	**88.34**
	√	√	**92.12**	**87.52**	**89.37**	90.78	85.49	87.53
GBC	√		73.79	78.38	75.28	76.25	77.52	75.93
		√	**89.28**	**86.53**	**86.78**	**91.51**	**87.90**	**89.26**
	√	√	87.01	86.27	86.28	86.15	85.33	85.19
	Feature		Data_ba3			Data_unba		
Classifier	T	O	Precision	Recall	F1	Precision	Recall	F1
XGB	√		72.19	73.19	71.97	66.56	41.31	50.70
		√	88.65	86.17	87.10	90.34	75.35	81.47
	√	√	**91.81**	**87.00**	**88.90**	**90.98**	**76.32**	**82.30**
RFC	√		75.17	74.63	74.16	75.85	38.08	50.29
		√	90.08	83.88	86.69	94.22	**73.70**	**82.23**
	√	√	**95.60**	**87.04**	**90.45**	**95.64**	72.40	81.86
GBC	√		72.90	75.90	73.65	64.10	35.23	45.03
		√	89.24	86.75	87.66	89.30	**74.36**	**80.46**
	√	√	**92.29**	**87.00**	**89.11**	**91.55**	72.81	80.22

Note: The bold entries represent the best results among the classifiers.

Obs. 4. Mixing multiple types of manual features does not mean better performance: In some cases, using both features can result in negative benefits compared to using opcode features alone. The possible reason is the similarity of the statistical transaction characteristics of the Ponzi and non-Ponzi accounts.

5.5.2. *Performance on network features*

We further investigate the effectiveness of network features on detecting Ponzi schemes. Specifically, we generate the embedding representations for contract accounts using three network embedding methods (DeepWalk, Node2vec, and LINE), and then feed the

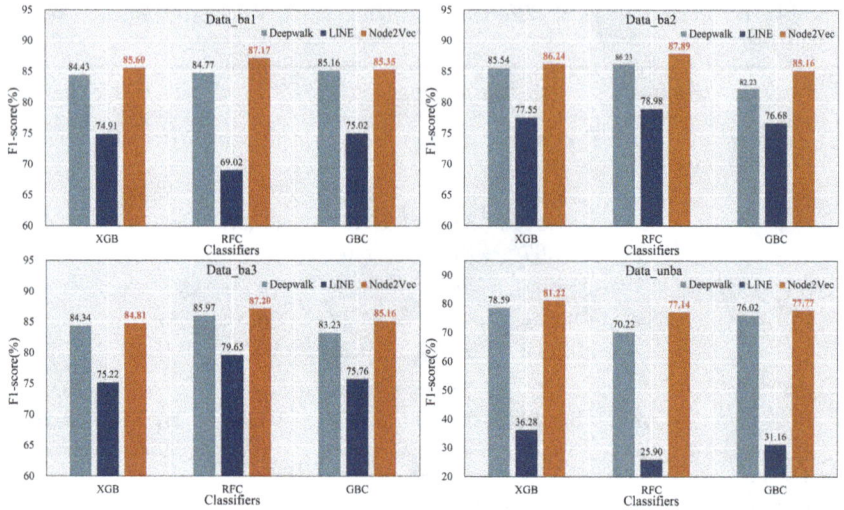

Fig. 7:   The performance of different embedding methods on the top three best classifiers.

learned network features (N) into the three ensemble learning classifiers (XGB, RFC, GBC). Figure 7 shows the performance comparison of network features extracted by three embedding methods. Since the structural features mentioned earlier are obtained by the embedding method, we use (E) to represent the structural features that follow.

Obs. 5. Effective network features benefit Ponzi scheme detection: We use three different graph embedding schemes in Fig. 7 to obtain the graph embedding features of the transaction network. Compared with the experimental data in Table 4, the network embedding feature performance outperforms the performance when using only manual transaction features. And the network embedding feature performance and the operational code features with outstanding performance are competitive to some extent. Among them, DeepWalk and node2vec perform better than LINE, and node2vec performs best. In a case of imbalanced datasets, RFC Ponzi scheme classifier with node2vec embedding method shows the best performance. XGB classification method outperforms the remaining two classification methods in the case of unbalanced datasets. We explore the reasons for this phenomenon. From the perspective of embedding methods, LINE performs poorly, suggesting that maintaining high-order proximity is not effective in distinguishing between Ponzi and non-Ponzi nodes. DeepWalk performs at a moderate level, slightly below node2vec. Compared to DeepWalk, node2vec has more selectivity because it can control the random walk strategy through the parameters p and q, where p is larger and q is smaller, prefer to visit a common neighbor, or visit other one-hop neighbors. They prefer the Depth-First Search (DFS), which can expresses Ponzi scheme homogeneity. p is smaller and q is larger, tend to visit the common domain and return to the intended node, preferring the Breadth-First Search (BFS) to express the Ponzi scheme structure with more possibilities. The BFS tends to search the Ponzi scheme structure. Therefore, the choice of node2vec can better express the centrality and similarity of proximity nodes, providing more potentially useful information for Ponzi scheme identification.

Obs. 6. Different embedding methods and dimensions affect Ponzi scheme identification: Next, we discuss the impact of embedding dimensions and classifiers. According to the results of the above experiments, we choose the best network embedding method, i.e., the node2vec embedding method based on random walk. Among them, we set the dimensions of network embedding to 16, 32, 64, and 128, respectively. In Fig. 8, from the perspective of the three optimal classifiers, the random forest performs best on the balanced dataset. XGB performed best in imbalanced dataset. From the perspective of different embedding dimensions, the implementation results show that different datasets have different optimal embedding dimensions, so it is difficult to compare in this respect. The following experiments all take the best dimension results.

5.5.3. *Performance on ALL features*

Further, we explore the classification effect of multiple feature combinations. We combine transaction features and embedded features into new features (TE), transaction features, opcode features, and embedded features into new features (TOE). The composition features are trained, and the following experimental results are obtained:

Obs. 7. Combined features exhibit optimal performance: In the balanced dataset, we use RFC as the best performing classifier to show

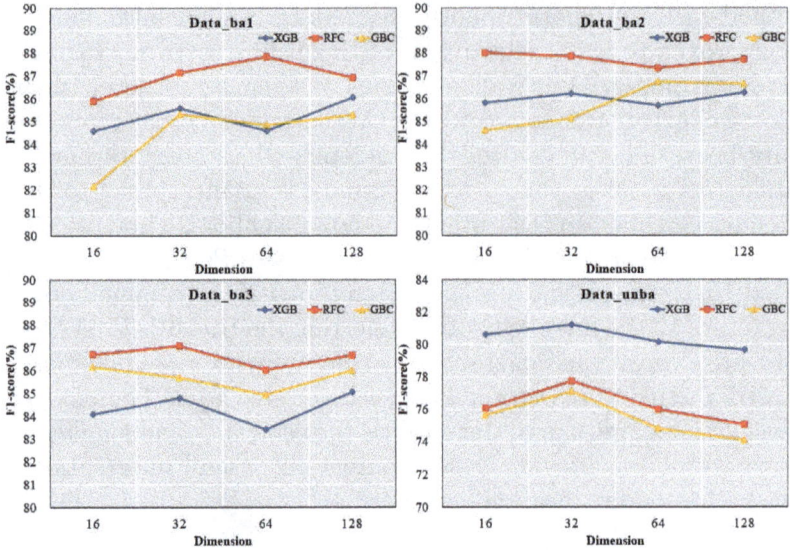

Fig. 8: The performance of node2vec with different embedding dimensions on the top three best classifiers.

the experimental results. In the imbalanced dataset, we use XBG as the best classifier to show the experimental Fig. 9. But no matter what kind of dataset, we found that when only using the embedded feature (E), the detection efficiency is better than the method of only the transaction feature (T), and the effect is similar to the performance of the unique opcode feature (O), so we can conclude that adding structural information can indeed learn more information representation, thereby improving the classification effect method. However, when the transaction feature is combined with the embedded feature (TE), the classification performance is inconsistent. With the performance of the embedded feature E alone, and performance on some datasets is reduced. Then the three features (TOE) are superimposed, and the classification effect is significantly better than other features.

Obs. 8. Better performance on graph deep learning than machine learning methods: On the other hand, we use an end-to-end GCN model, and only need to input features and network structure to obtain the recognition result of the Ponzi contract. First of all, the network constructed by the imbalanced dataset is too large to use

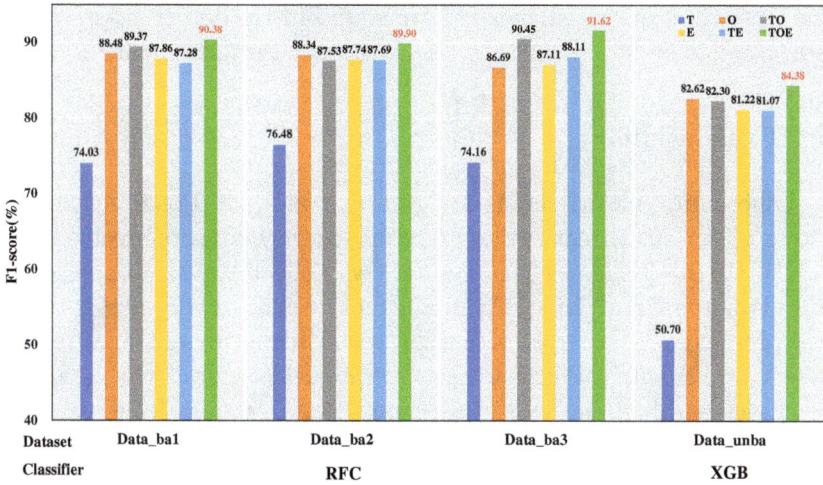

Fig. 9:   The performance of all the features of the imbalanced dataset on the eXtreme Gradient Boosting.

Table 5:   Performance of the GCN model and the RFC using T and TO features.

Features	Methods	Data_ba1			Data_ba2			Data_ba3		
		Pre	Recall	F1	Pre	Recall	F1	Pre	Recall	F1
T	RF	75.32	75.51	74.03	79.69	76.83	76.48	75.17	74.63	74.16
	GCN	**82.76**	**77.89**	**79.74**	**83.33**	**76.32**	**79.28**	**86.57**	**77.89**	**81.60**
TO	RF	85.59	81.05	82.64	89.36	80.53	84.27	88.28	79.47	83.07
	GCN	**90.05**	**87.50**	**88.31**	**90.47**	**85.70**	**87.42**	**95.60**	**87.04**	**90.45**

Note: The bold terms are the best performances compare to that of other methods.

GCN experiments, so we only use the balanced dataset for experiments. Secondly, we use the earlier-mentioned classifier with the best performance, the random forest classifier, as the baseline. Finally, we compare the embedding vectors of different dimensions of GCN, and the experiment finds that when the dimension of the embedding vector is 128, the classification effect is the best. The final experiment results are shown in the following Table 5.

We find that the precision is significantly improved by using the GCN approach in all cases. The improvement over the RFC baseline is 3–6% when using only transaction and network structure features as inputs. Further, after introducing the opcode features, not only is the F1 of the baseline obviously improved, but the detection precision is also up to 90.45% using the GCN approach in Data_ba3.

6. Conclusion

In Ethereum, various scams are rampant and the Ponzi scheme is the most serious threat to the financial security of users involved. To deal with this issue, in this chapter, we collect transaction records from Ethereum and detect Ponzi schemes from a network perspective. We propose a GCN model that incorporates extracted features to detect and identify Ponzi smart contracts. Compared with general machine learning methods, the experimental results indicate that our proposed model performs better than a series of baselines to identify the Ponzi schemes. In the future, we plan to further extend our proposed model to detect Ponzi schemes in time.

References

1. V. Buterin *et al.*, A next-generation smart contract and decentralized application platform, *White Paper.* **3**(37) (2014).
2. G. Wood *et al.*, Ethereum: A secure decentralised generalised transaction ledger, *Ethereum Project Yellow Paper.* **151**(2014), 1–32 (2014).
3. K. Wu, An empirical study of blockchain-based decentralized applications, arXiv preprint arXiv:1902.04969 (2019).
4. M. Bartoletti, S. Carta, T. Cimoli, and R. Saia, Dissecting ponzi schemes on ethereum: Identification, analysis, and impact, *Future Generation Computer Systems.* **102**, 259–277 (2020).
5. Y. Wenqiang, Z. Yanmei, L. Ziyu, and N. Wa, Type analysis and identification method of Ethereum Ponzi scheme, *Journal of Chongqing University.* **43**(11), 111–120 (2020).
6. Y. Lou, Y. Zhang, and S. Chen, Ponzi contracts detection based on improved convolutional neural network, in *2020 IEEE International Conference on Services Computing (SCC)*, 2020, pp. 353–360.
7. E. Jung, M. Le Tilly, A. Gehani, and Y. Ge, Data mining-based ethereum fraud detection, in *2019 IEEE International Conference on Blockchain (Blockchain)*, 2019, pp. 266–273.
8. T. Hu, X. Liu, T. Chen, X. Zhang, X. Huang, W. Niu, J. Lu, K. Zhou, and Y. Liu, Transaction-based classification and detection approach for ethereum smart contract, *Information Processing & Management.* **58** (2), 102462 (2021).

9. L. Wang, H. Cheng, Z. Zheng, A. Yang, and X. Zhu, Ponzi scheme detection via oversampling-based long short-term memory for smart contracts, *Knowledge-Based Systems.* **228**, 107312 (2021).

10. W. Chen, Z. Zheng, J. Cui, E. Ngai, P. Zheng, and Y. Zhou, Detecting ponzi schemes on ethereum: Towards healthier blockchain technology, in *Proceedings of the 2018 World Wide Web Conference*, 2018, pp. 1409–1418.

11. W. Chen, Z. Zheng, E. C.-H. Ngai, P. Zheng, and Y. Zhou, Exploiting blockchain data to detect smart ponzi schemes on Ethereum, *IEEE Access.* **7**, 37575–37586 (2019).

12. Y. Zhang, W. Yu, Z. Li, S. Raza, and H. Cao, Detecting Ethereum Ponzi schemes based on improved LightGBM algorithm, *IEEE Transactions on Computational Social Systems.* **9**, 624–637 (2021).

13. Zhang Yan-mei and Lou Yi-cheng, Deep neural network based Ponzi scheme contract detection method, *Computer Science.* **48**(1), 273 (2021). doi: 10.11896/jsjkx.191100020.

14. S. Fan, S. Fu, H. Xu, and C. Zhu, Expose your mask: Smart Ponzi schemes detection on blockchain, in *2020 International Joint Conference on Neural Networks (IJCNN)*, 2020, pp. 1–7.

15. J. Wu, J. Liu, Y. Zhao, and Z. Zheng, Analysis of cryptocurrency transactions from a network perspective: An overview, *Journal of Network and Computer Applications.* **190**, 103139 (2021).

Chapter 12

Research on Prediction of Molecular Biological Activity Based on Graph Convolution

Yinzuo Zhou*, Lulu Tan, Xinxin Zhang, and Shiyue Zhao

Alibaba Business School, Hangzhou Normal University,
Hangzhou, P.R. China
Correspondence School, Hangzhou Dianzi University,
Hangzhou, P.R. China

*zhouyinzuo@163.com

The drug development cycle is long and expensive. Using computer algorithms to screen lead compounds can effectively improve its efficiency. Moreover, the quantitative structure–activity relationship modeling methods can be used to predict the biological activity of molecules. This has become a major research focus in the field of drug development. However, due to the limitation of methods and computing power, the existing machine learning-based modeling techniques cannot meet the requirement of the big data-driven drug development. The purpose of this chapter is to construct a more reliable prediction model for molecular biological activities. The instability and unreliability caused by artificial computing features are avoided by learning molecular graph features directly. During the process of modeling, we address problems such as adaptive learning in feature fusion and sample balance, thus improving the overall performance of the model.

1. Introduction

The drug development cycle is long and the cost is high. The sudden decrease of drugs in clinical research can lead to huge waste of resources. At present, 9 of every 10 candidate drugs fail in phase-I clinical trials or regulatory approval.[1] To improve the low efficiency of the drug development process, we aim to shorten the cycle of new drug research and improve the success rate of the drug development. Pharmaceutical chemists give the concept of quantitative structure–activity relationships (QSAR). QSAR aims to quantitatively determine the biological activity of a series of derivatives of known lead compounds, analyze the relationship between the main physical and chemical parameters of derivatives and biological activity, and establish a mathematical model between structure and biological activity to guide drug molecular design.[2] Machine learning methods are common for chemical informatics. Since traditional machine learning methods can only deal with fixed-size inputs, most early QSAR modeling use manually generated molecular descriptors for different tasks. Common molecular descriptors include: (1) molecular fingerprints, which encode the molecular structure through a series of binary numbers representing the specific substructure;[2] (2) one/two-dimensional molecular descriptors, which are derived from molecular physical chemistry and differential topology processed by statisticians and chemists.[3] Common modeling methods include linear methods (such as linear regression) and nonlinear methods (such as support vector machine and random forest, etc.). Recently, deep learning has become the leading technique for QSAR modeling.

In the past decade, deep learning has become a popular modeling method in various fields, especially in medicine. It is used in biological activities and physicochemical properties' prediction, drug discovery, medical image analysis, and synthetic prediction, etc. Convolutional neural networks (CNN) are special in deep learning, and have successfully solved many problems with structured data (such as images).[4] However, graph data have irregular shape and size; node position has no spatial order; and the neighbor of the node is also not related to the position. Hence, the traditional CNN cannot be directly applied to the graph. For the non-Euclidean structure data, graph convolutional network (GCN) is proposed, along with various derivative architectures. In 2005, Gori *et al.*[5] proposed the first graph neural network (GNN), which learns the architecture of undirected graphs, directed graphs, and cyclic graphs based on recurrent neural networks. In 2013, Bruna *et al.*[6] proposed GCN based on spectral graph theory. There are other forms of GCN, such as graph attention network (GAT),[7] graph automatic encoder,[8] and spatiotemporal graph volume.[9]

Recently, researchers have applied GCN to molecular bioactivity prediction. In chemical graph theory, the structure of compounds is usually represented as a molecular graph of hydrogen depletion (omitting hydrogen). Each compound is represented by an undirected graph, with atoms as nodes and bonds as edges. Both atoms and bonds contain many attributes, such as atomic type, bond type, and so on. In 2016, Kearnes *et al.*[10] established a graph convolution model using the attributes of nodes (atoms) and edges (bonds). In 2017, Connor *et al.*[11] created atomic eigenvector and bond eigenvector and spliced them to form atomic bond eigenvector. In 2018, Pham *et al.*[12] proposed the graph memory network (GraphMem), a memory-enhanced neural network, which can be used to process molecular graphs with multiple bond types. In these studies, node characteristics and bond attributes are not distinguished. Their internal relations are not analyzed. However, giving different weights to various interaction types between atomic pairs is more accurate.

Shang *et al.*[13] proposed an edge attention graph convolutional network (EAGCN) algorithm based on edge attention. The algorithm proposed an edge attention layer to evaluate the weight of each edge in the molecule. An attribute tensor is constructed in advance and multiple attention weight tensors are generated after being processed by the attention layer. Each of them contains all possible attention weights of an edge attribute in the dataset (molecular graph). Then the attention matrix is constructed by finding the value of each bond of the molecule in the weight tensor. This method enables the model to learn different attention weights at different levels and different edge attributes. Experiments show that EAGCN framework has high applicability, and can learn specific molecular features directly from the graph structure. Therefore, it can avoid the error caused by the data preprocessing stage.

Considering inability of the EAGCN framework, to adaptively learn the features, we propose an attention graph convolution model based on multi-feature fusion. The multi-feature fusion scheme is a feature fusion method based on self-attention mechanism, which can effectively make the model adaptively adjust the weight distribution of multiple feature tensors. In this chapter, we use a variety of screening methods[14] to limit the targets in the PubChem database and validate the effectiveness of our method on several different types of bioactive datasets. Compared with various baseline models, the experimental results show that our method can achieve better performance.

We use different types of datasets. The first dataset is from PubChem, a public chemical database. In addition, we use four datasets of cytochrome P450 series from the 1851 target family, two inhibitor activity datasets, and a molecular set for the recognition of the binding R (CAG) RNA repeat sequence.

Moreover, a GCN architecture based on edge attention is applied to the biological activity dataset selected to learn molecular graphs directly and avoid errors caused by artificial feature engineering. This model shows better classification performance than traditional machine learning methods, and its accuracy index can be 2–8% higher. We also research the feature fusion method and propose a multi-feature fusion scheme. Since the existing models are unable to adaptively learn the edge attribute feature weight, we propose a molecular multi-feature fusion scheme to optimize the feature extraction ability of the algorithm model. Adaptive fusion for multiple features is realized through the attention mechanism. The accuracy index can be improved by 1–2%.

We study the problem of sample imbalance and propose a loss optimization plan. To address the imbalance of positive and negative samples and difficult samples in molecular biological activity data, the loss calculation scheme was improved. Two novel loss modification schemes, namely, focal loss and gradient harmonizing mechanism (GHM), are introduced to further optimize the model performance.

2. Method

2.1. *Establishment of graphs*

In chemical graph theory, the structure of compounds is usually represented as a molecular graph of hydrogen depletion (omitting

Table 1: Atomic attribute expression.

Atomic properties	Description	Value type
Atomic number	The position of atoms in the periodic table	Int
Number of connected atoms	Number of neighbor nodes	Int
Number of adjacent hydrogen atoms	Number of hydrogen atoms	Int
Aromaticity	Aromatic or not	Boolean
Number of formal charges	Number of formal charges in the ring or not	Int
Ring state		Boolean

Table 2: Bond attribute expression.

Bond properties	Description	Value type
Atomic pair type	Atomic type definition of bond connection	Int
Bond order	Single bond/double bond/triple bond/aromatic bond	Int
Aromaticity	Aromatic or not	Int
Conjugation	Conjugate or not	Boolean
Ring state	In the ring or not	Boolean
Placeholder	There is a bond between atoms or not	Boolean

hydrogen). Each compound can be represented by an undirected graph with atoms as nodes and bonds as edges. Among them, the molecular attribute information includes atomic attribute and bond attribute[14] as shown in Tables 1 and 2. These properties are very important to describe the bonding strength, atom activities, and bonding resonance between two atoms. Different edge attributes are treated as attention layers and correspond to different edge attention matrices.

The following definitions are applied.

(1) **Definition 1**: Given a graph $G = (V, E)$, V is a finite set of nodes; $N = |V|$ is the number of nodes; and $E \subseteq V \times V$ is a finite set of edges. The adjacency matrix A of G is defined as a square matrix

with dimension $N \times N$. $a_{ij} = 1$ means that there is a connected edge between nodes i and j. Otherwise, $a_{ij} = 0$.

(2) **Definition 2**: The node feature matrix of G can be denoted as $H^l \in R^{N \times F}$, where F is the total number of features of each node. Its line represents the characteristics of node i and a series of edge attributes. A molecular attribute tensor $M \in R^{N_{\text{atom}} \times N_{\text{atom}} \times N_{\text{features}}}$ (i.e., N_{features} is the F) for G is the input of the model.

2.2. Graph convolution based on edge attention

EAGCN[13] learns different attention weights at different levels and edge attributes to construct a molecular attention matrix. The algorithm constructs an attribute tensor in advance. After attention layer processing, multiple attention weight tensors are generated, each of which contains all possible attention weights of an edge attribute in the dataset. Then the attention matrix is constructed by finding the value of each bond of the molecule in the weight tensor. This method allows different molecules to correspond to different attention matrices.

EAGCN uses the atomic and bond properties of molecules to construct an adjacency matrix A, a node characteristic matrix H^l, and a molecular attribute tensor M for each molecule. The general process of the model is shown in Fig. 1. The whole model takes the molecular graph as the input. After processing the edge attributes in the molecular graph, the edge attribute tensor is finished. After one-hot coding, five graph convolution features are obtained through the GAT layer, respectively. Then the total tensor features are obtained through splicing, which is used as the input of the next GAT layer. Finally, the two dense layers are used to output the results.

(1) Attention Layer $A_{att,i}$ Construction

After the molecular graph attribute feature selection, the adjacency matrix A and molecular attribute tensor M can be obtained. In order to calculate the weight of different edges in each attribute, we split the molecular attribute tensor M according to K edge attributes (we select the first five edge attributes in Table 2, i.e., $K = 5$). Since each edge attribute contains different state values, one-hot coding generates five adjacency tensors $T_i^l \in R^{N * N * d_i}$ with dimension $N * N * d_i$, where d_i is the number

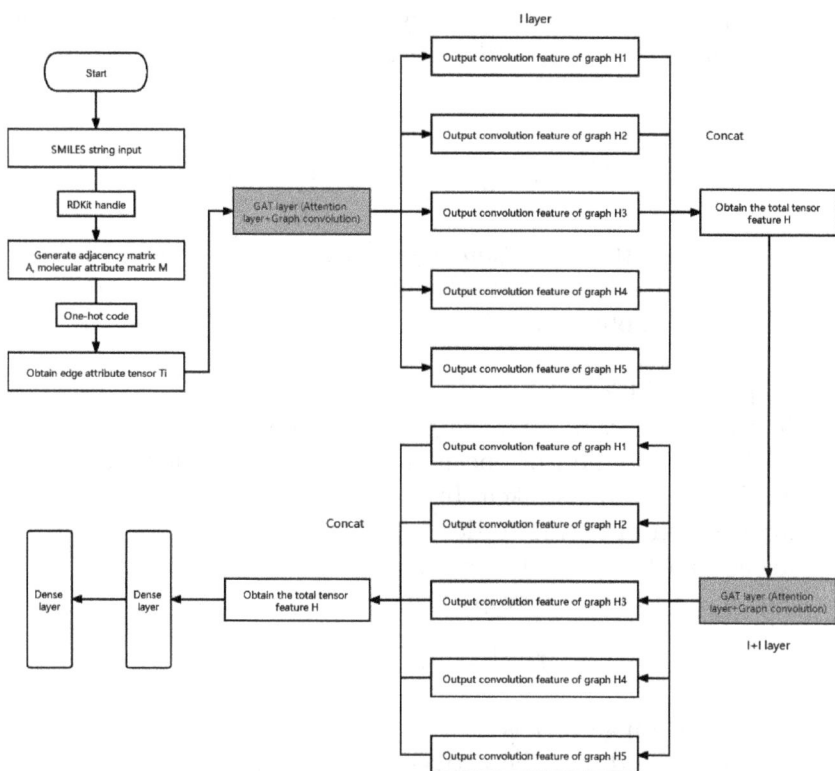

Fig. 1: Flow chart of Edge Attention-based Graph Convolution Network (EAGCN) model.

of edge attribute types. Through the edge attribute, tensor T_i can get the weight $a_{i,j}^l$ in $A_{att,i}^l$ (shown in Eq. (1))

$$A_{att,i}^l = < T_i^l, \quad D_i^l >$$ (1)

$T_i^l \rightarrow A_{att,i}^l$ is processed as follows.

(i) Firstly, a filter D_i^l with size of $1 * 1 * d_i$ is used through convolution processing with d_i input channels and one output channel. It moves with steps 1, which represents the attention layer at the edge of layer l.

(ii) Secondly, in order to make the weights comparable in different edges, the softmax function is used to normalize the weights (Eq. (2)). The output values obtained by the softmax function are interrelated. It can quantify them in the range of 0–1, and the sum of the output values is 1.

$$\left(\widetilde{A}_{\text{att}},i\right)_{s,t} = \frac{\exp\left(A^l_{\text{att},i}\right)_{s,t}}{\sum_{t=1}^{M}\exp\left(A^l_{\text{att},i}\right)_{s,t}}. \tag{2}$$

In general, the attention layer passes the two-dimensional (kernel_size is 1×1) and softmax result, for attribute adjacency tensor T_i^l, to obtain weight matrix of an edge $A^l_{\text{att},i}$ for attribute of each edge.

(2) Map Convolution

In an adjacency matrix, the graph volume product only focuses on the information of one node and adjacent nodes, i.e., only local information is taken. In each graph volume layer, the node information of all first-order neighbors is aggregated. Then the linear transformation is carried out. We calculate the convolution characteristics of the graph using Eq. (3):

$$H_i^{l+1} = \sigma\left(\widetilde{A}_{\text{att},i}H^l W_i^l\right), \tag{3}$$

where the range of i is $1 \leq i \leq K$, K is the number of edge attributes, and σ is the activation function (ReLU). Each edge attribute i generates a value in the $\widetilde{A}_{\text{att},i}$. Therefore, $\widetilde{A}_{\text{att},i}H^l$ can be regarded as the weighted sum of node features. Next, the graph convolution features obtained from different attribute tensors are spliced. Concat is directly used to form the total feature tensor $H^{l+1} = \left\{H_i^{l+1} \in R^N \times R^{F_i'} \mid 1 \leq i \leq K\right\}$.

The algorithm repeats the above steps and extracts the graph convolution feature of the next layer with H^{l+1}. After two layers of processing by the GAT layer, two fully connected layers are connected to calculate the classification confidence of the molecular model.

2.3. Attention graph convolution based on multi-feature fusion

EAGCN is evaluated in different kinds of bioactivity prediction datasets. The model outperforms the traditional machine learning methods. The performance of EAGCN model can be attributed to the following characteristics:

(1) By directly learning the molecular graph, it can avoid the error caused by manual screening features and its impact on the robustness and reliability of the model.
(2) The generated attention weight matrix depends on the domain characteristics of a node instead of global characteristics. The weight is shared in all graphs. Therefore, the local characteristics of the extracted data can be realized through the shared features.

However, when the molecular attribute matrix is used as the input of the attention layer (including molecular and edge information), EAGCN uses the attention layer to process the attribute information and obtain the weight of each edge, While in the feature fusion, the concat method only performs simple dimension splicing, which leads to the non-discrimination of multiple attribute information. Adding dimensions may also reduce the model computational efficiency and affect the performance of models. Therefore, we propose a multi-feature fusion scheme based on attention, which can further use the edge attribute information of molecular graph for feature extraction and improve the overall performance.

In the original model, after the weight tensor gets the features through graph convolution, concat is used to merge the channels and integrate the feature graph information. Therefore, based on the feature fusion scheme of self-attention mechanism, we set a higher channel number for the characteristic matrix of atomic pair type, which is equivalent to manually setting bias weight.

Thus, we propose a multi-feature fusion method for algorithm optimization. This is a feature fusion scheme based on self-attention mechanism,[15] which can give different weight parameters to each input element to pick out the more important information in each feature and suppress but not lose other information. It can process the global connection and local connection simultaneously, which can further improve the learning efficiency of the model.

2.3.1. *Multi-Feature Fusion Scheme*

EAGCN generates the molecular attribute tensor $M \in R^{N_{atom} \times N_{atom} \times N_{features}}$ for each graph. In order to calculate the

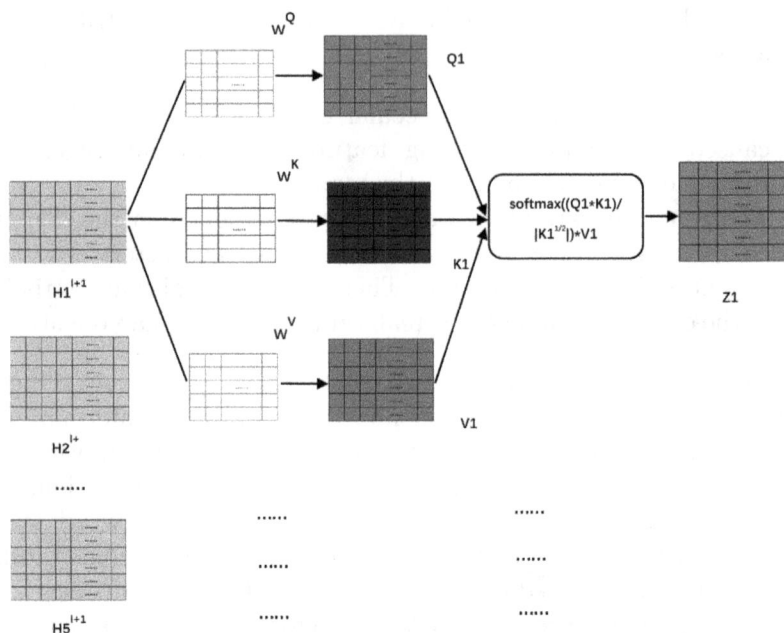

Fig. 2:　Flow chart of multi-feature fusion model.

weights of different edges in each attribute, we encode the molecular attribute tensor M as one-hot. Then multiple attribute tensors are input into the attention layer to obtain weight matrices $A_{\text{att},i}^l$. After the convolution of the graph, the graph convolution characteristic matrix H^{l+1} is obtained. Then we replace concat fusion method with multi-characteristic fusion scheme. The specific steps are as follows:

(1) The algorithm generates Q, K, and V weight tensors for each input. After the graph convolution feature H^{l+1} is obtained, the five graph feature tensors are used as inputs. The dimension of H^{l+1} varies according to the number of channels set in the model. Let's take a graph characteristic matrix with dimension $N * 30$ as an example. First, we set three different weight tensors for each characteristic tensor, namely, query Q, bond K, and value V tensors. The default length is 64. These three tensors are obtained by multiplying the input features and the weight matrix. The calculation example is shown in Fig. 2. W^Q, W^K, W^V are three different weight matrices with same dimensions (30×64).

The characteristic tensor H_i^{l+1} is multiplied by them to obtain the corresponding Q, K, and V tensors.

(2) The score is calculated. The bond vector and query vector of each feature are multiplied to obtain $score = Q \bullet K$ ($\bullet$ represents the dot product operation in a matrix).

(3) In order to make the gradient more stable, we divide the score calculated in step 2 by $\sqrt{d_k}$ and normalize it. d_k is the module length of the K vector, i.e., 64.

(4) The result is normalized using Softmax. Softmax is used to normalize the scores of all characteristic tensors so that the obtained scores are positive and the sum is 1. The purpose of this step is to preliminarily obtain the weight of each edge attribute for the whole graph.

(5) Multiply the value vector V by the Softmax score points to obtain the score V of each weighted input tensor (graph convolution feature).

(6) Sum the weighted value vector to obtain the output $z = \sum v$, $z \in R^N * R^{30}$, which is the weight matrix z.

2.4. Datasets

The datasets are from a public Chemistry Database PubChem.[16] We selected a variety of analysis and screening methods in the literature[14] and different types of biological activity datasets. We also limited the screening targets, such as screening multiple series of cytochrome P450 enzymes. Finally, four datasets of cytochrome P450 Series in 1851 target family, two inhibitors, and molecular series identifying binding r (CAG) RNA repeats were selected. Table 3 lists the relevant information and filtering conditions of the selected dataset.

3. Experimental Results

3.1. Data processing

In the study of molecular bioactivity, the input data is the basis of QSAR research. The forms of molecular input data used by different algorithm models are also different. The common representations of molecules are molecular identifier and molecular descriptor.

Table 3: Classification dataset information from PubChem database used in this Chapter.

PubChem AID	Screening conditions	Number of active molecules	Number of inactive molecules
1851(1a2)	Cytochrome P450, family 1, subfamily A, polypeptide 2	5,997	7,242
1851(2c19)	Cytochrome P450, family 2, subfamily C, polypeptide 19	5,905	7,522
1851(2d6)	Cytochrome P450, family 2, subfamily D, polypeptide 6, isoform 2	2,769	11,127
1851(3a4)	Cytochrome P450, family 3, subfamily A, polypeptide 4	5,265	7,732
492992	Identify inhibitors of the two-pore domain potassium channel (KCNK9)	2,097	2,820
492992	Inihibition of *Trypanosoma cruzi*	4,051	1,326
652065	Identify molecules that bind r (CAG) RNA repeats	2,969	1,288

(1) *Molecular identifier*: There are text-based identifiers such as simplified molecular input line entry system (SMILES)[17] and international chemical identifier (InChI).[18] SMILES is a text string encoding three-dimensional chemical structure with a set of ordered rules and special syntax. It is a language structure used to store chemical information. For example, the SMILES identifier of carbon dioxide CO_2 is O = C = O. SMILES is an identifier commonly used in QSAR modeling. The international chemical identifier InChIs is not often used in deep learning because of the instability of prediction performance.

(2) *Molecular descriptors*: Molecular descriptors are the basis of early QSAR research. The traditional machine learning model cannot identify and deal with the molecular structure. The physical and chemical properties or structure-related parameters of molecules are derived using various algorithms.

At present, there are many computing tools for molecular descriptors, including various open sources or commercial software and libraries. Recently, commonly used molecular descriptor calculation software has included Dragon,[19] alvaDesc,[20]

Gaussian,[21] Padel-Descriptor,[22] OpenBabel,[23] etc. Among them, the classic Dragon software can calculate thousands of molecular descriptors. Chemical libraries include RDkit.[24] It can calculate various molecular descriptors and carry out molecular visualization and chemical analysis.

MF_EAGCN and EAGCN, Random Forest (RF), Support Vector Machines (SVM), and Deep Neural Networks (DNN) are compared. For the traditional machine learning methods (RF, SVM, DNN), the calculated molecular descriptors need to be used. Therefore, for the molecular SMILES data, 200 one-dimensional molecular descriptors generated by RDKit (open source chemical computing software package) are used as the features of the benchmark model. At the same time, the atomic and edge properties of molecules calculated by RDKit are also used.

3.2. *Experimental setup*

We first apply EAGCN to different types of biological activity classification datasets. Then we apply the attention map volume based on multi-attribute fusion to the same dataset. The purpose of the experiment is to: (1) verify that the graph convolution model based on edge attention can indeed improve the classification performance of bioactive data compared with traditional machine learning methods (such as random forest, deep neural network, etc.); (2) verify the performance improvement of the attention graph convolution model based on multi-feature fusion in the biological activity prediction task; and (3) verify the problems still existing in the model, and design the following optimization scheme.

The K-fold cross-validation method is used for the dataset division. 8-fold cross-validation method is used for dataset partition in this chapter. Different random seeds are used three times. Similarly, the results obtained here are the average of three runs and the standard deviation is listed.

(1) *Benchmark Methods*: The benchmark methods used in this chapter are random forest (RF), support vector machine (SVM), and deep neural network (DNN) as shown in Table 4. The Hyperparameter list is set for model parameter adjustment. Similarly, the

dataset is divided by the 20% cross-validation method and then executed three times with different random seeds. The results are the average of three runs and the standard deviation is recorded.

(2) Experimental Setup of EAGCN and MF_EAGCN: During EAGCN modeling, the model performance is better when the weight of the atomic pair-type attribute is large. Therefore, in this algorithm, we manually set the number of output channels of the atomic pair-type GCN layer to be large. In our optimized MF_EAGCN, we pay attention to the edge attributes with higher weight. Thus, we can adaptively learn different edge attribute weights. The experimental parameters set in this chapter are shown in Table 5.

(3) Evaluation indicators: We use two evaluation indicators: accuracy (ACC) and balanced score ($F1$-score). Accuracy (ACC) is a commonly used evaluation index in classification prediction (Eq. (4)).

$$\text{accuracy} = (\text{TP} + \text{TN})/(\text{P} + \text{N}), \tag{4}$$

Table 4: Hyperparameter setting of each model.

Hyperparameter	Value range	Parameter meaning
Random forest		
Ntrees	(50, 100, 150, ..., 500)	Number of trees
max_depth	(1, 5, 10, ..., 45, 50)	Maximum depth per tree
max_features	(1, 5, 10, ..., 45, 50)	Maximum characteristic number of partition time
Support vector machines		
Kernel	RBF	Kernel function
C	(1, 10, 100)	Penalty coefficient
γ	(0.1, 0.001, 0.0001, 0.00001, 1, 10, 100)	Affects the amount of data mapped to the new feature space
Deep neural networks		
Epoch	100	Number of iterations
Batch size	100	Minimum number of training samples
Hidden layers	(2, 3, 4)	Number of hidden layers
Number neurons	(10, 50, 100, 500, 700, 1000)	Number of neurons per layer
Loss function	ReLu	Neuron activation function
Aromaticity	binary_crossentropy	Loss function

Table 5: EAGCN and MF_EAGCN model Hyperparameter setting table.

Hyperparameter	Value range	Parameter meaning
EAGCN		
Batch size	64	Number of single training samples
Epoch	100	Number of iterations
weight_decay	0.00001	Weight decay rate
dropout	0.5	Random inactivation rate
Activation function	ReLu	Activation function
Loss function	binary_crossentropy	Loss function
kernel_size	1	Convolution kernel size
stride	1	Convolution kernel sliding step
n_sgcn1	(30, 10, 10, 10, 10)	Number of convolution output channels of multi feature map
MF_EAGCN		
Batch size	64	Number of single training samples
Epoch	100	Number of iterations
weight_decay	0.00001	Weight decay rate
dropout	0.5	Random inactivation rate
Activation function	ReLu	Activation function
Loss function	binary_crossentropy	Loss function
kernel_size	1	Convolution kernel size
stride	1	Convolution kernel sliding step
n_sgcn1	(20, 20, 20, 20, 20)	Number of convolution output channels of multi-feature map

where TP and TN are the numbers correctly classified into positive and negative cases, respectively; and P and N are the number of positive and negative cases in the actual sample. Higher ACC value indicates better classifier performance.

The balanced $F1$-score is also an indicator commonly used to measure the accuracy of the model in biological activity classification tasks (Eq. (5)).

$$F_1 = 2 \cdot \frac{\text{precision} \cdot \text{recall}}{\text{precision} + \text{recall}}. \tag{5}$$

$F1$-score takes into account both the precision and recall of the model. Only when both values are high, the value of $F1$ is higher and the model performance is better.

3.3. Algorithm performance analysis

Table 6 shows the ACC and $F1$-score index results of different benchmark models on several datasets.

EAGCN based on graph convolution shows better classification performance than traditional machine learning methods. Its ACC index is 2–8% higher than the benchmark learning model. Its $F1$-score index is 1–5% higher than the benchmark learning model. The information obtained directly from the molecular graph rather than the pre-calculated characteristics can improve the model performance. In a few datasets, the performance of DNN is basically the same as or slightly higher than that of EAGCN. The performance of RF can sometimes be the same as EAGCN. There is still plenty of room for optimization. Based on multi-feature fusion, MF_EAGCN model shows better classification performance. This proves that the multi-feature fusion scheme can make more use of edge attribute information to improve the model prediction performance. The ACC index is 1–2% higher, and the $F1$-score index is about 1% higher, than that of EAGCN model.

Figures 3 and 4 show the proposed MF_EAGCN. Benchmark algorithms EAGCN and traditional machine learning methods are applied to compare the distribution of ACC index and $F1$-score index in seven biological activity datasets. The entries of the histogram are RF, SVM, DNN, EAGCN, and MF_EAGCN model from left to right. In the ACC index distribution diagram, the effect of dataset 1851(2d6) on EAGCN model is not significant. There are two reasons. First, the amount of data is larger and the distribution of feature importance is uneven in the feature fusion stage of the model, which results in the neglect of important information. It can reduce the model prediction performance. Second, the ratio of positive and negative samples is 1:5, which is imbalanced and can also limit the improvement of model performance.

Table 6: Prediction results of seven datasets under this algorithm, EAGCN and three benchmark methods.

Task	ACC					F1-score				
Dataset	RF	SVM	DNN	EAGCN	EAGCN_MF	RF	SVM	DNN	EAGCN	EAGCN_MF
1851(1a2)	0.824 ±0.005	0.8 ±0.02	0.835 ±0.015	**0.85** ±**0.01**	**0.859** ±**0.012**	0.792 ±0.01	0.78 ±0.008	0.8 ±0.007	**0.83** ±**0.012**	**0.841** ±**0.01**
1851(2c19)	0.776 ±0.01	0.75 ±0.009	0.79 ±0.002	**0.802** ±**0.007**	**0.815** ±**0.003**	0.8 ±0.004	0.77 ±0.005	0.823 ±0.01	**0.84** ±**0.01**	**0.852** ±**0.008**
1851(2d6)	**0.849** ±**0.006**	0.83 ±0.007	0.84 ±0.002	0.843 ±0.005	**0.851** ±**0.003**	0.828 ±0.013	0.8 ±0.004	0.82 ±0.003	**0.83** ±**0.01**	**0.834** ±**0.006**
1851(3a4)	0.77 ±0.006	0.737 ±0.004	0.792 ±0.008	**0.817** ±**0.006**	**0.825** ±**0.01**	0.73 ±0.003	0.701 ±0.006	0.74 ±0.01	**0.791** ±**0.008**	**0.807** ±**0.005**
492992	0.713 ±0.004	0.705 ±0.006	0.745 ±0.005	**0.757** ±**0.01**	**0.762** ±**0.01**	0.683 ±0.005	0.674 ±0.006	0.692 ±0.009	**0.74** ±**0.01**	**0.75** ±**0.009**
651739	0.753 ±0.004	0.753 ±0.006	0.814 ±0.014	**0.83** ±**0.006**	**0.843** ±**0.003**	0.8 ±0.003	0.776 ±0.009	0.88 ±0.006	**0.882** ±**0.007**	**0.891** ±**0.002**
652065	0.75 ±0.004	0.7 ±0.005	0.755 ±0.015	**0.77** ±**0.006**	**0.774** ±**0.005**	0.73 ±0.008	0.67 ±0.009	**0.796** ±**0.012**	0.787 ±0.01	**0.792** ±**0.01**

Note: The bold entries represent the best values.

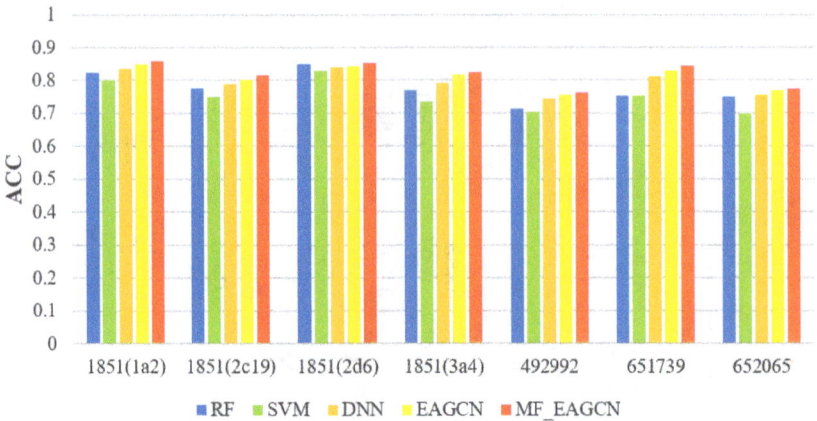

Fig. 3: ACC index distribution used to represent the performance of seven bioactive datasets in five classifiers.

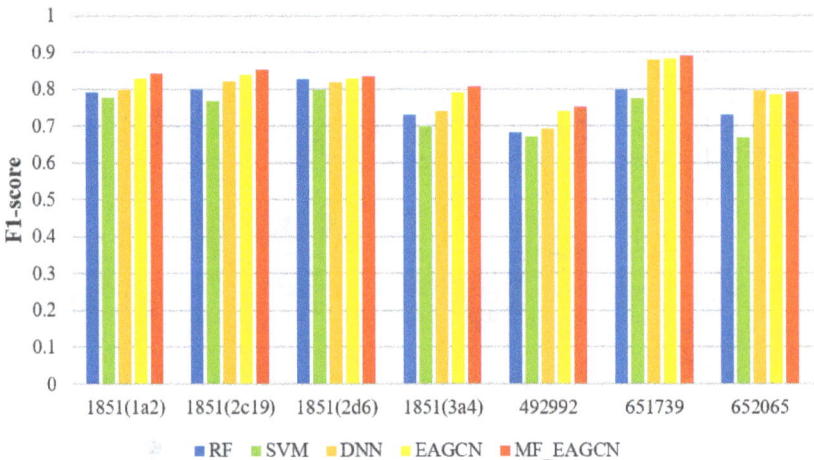

Fig. 4: F1-score index distribution used to represent the performance of seven bioactive datasets in five classifiers.

4. Discussions and Optimization

4.1. *Problem analysis*

The imbalance problem of difficult and easy samples is different from the imbalance problem of the number of positive and negative samples.[25] For general classifiers, classifying simple samples has little significance for the training model. Therefore, this part of parameter updating does not improve the ability of the model.

There are two main perspectives to solve this sample imbalance problem: data and algorithm.

(1) From the perspective of data, we try to modify and balance the distribution of various categories.[26] The most common methods are expanding the dataset and balanced sampling.[26] Sampling mainly includes oversampling and undersampling.[27]

(2) From the perspective of algorithm, we modify the algorithm model or loss function. Filtering out a large number of large and simple samples and increasing the proportion of loss in the overall loss value can alleviate the imbalance problem, which is also the starting point of this chapter.[28]

Sample category imbalance is a common and very important problem in many interdisciplinary fields. When the data are imbalanced, it is easy to reduce the model performance. This classifier cannot solve practical problems and cannot really realize data mining. Therefore, we propose two optimization schemes for the sample imbalance problem in this algorithm.

4.2. Model optimization

The prediction ability of the proposed algorithm is accurate. However, the imbalance of easy and difficult sample categories of bioactive data is still the main factor affecting its accuracy. Therefore, we propose to solve this problem by reducing weight of the loss function for the easy samples. The idea is to focus the training on fewer difficult samples. In order to evaluate the effectiveness of the optimization scheme, we perform experiments using different loss functions based on the MF_EAGCN model. The results show that the model can obtain better performance after optimizing the loss function.

4.2.1. Focal Loss

Focal Loss[29] is developed to solve the problem of imbalance in the number of difficult and easy samples. During the model training, a large number of samples are classified as easy samples. We should also consider the balance of difficult and easy samples. The loss of individual sample is very low. However, due to the extremely inbalanced number, the impact of easy samples is large, which eventually dominates the total loss. Therefore, if the weight of simple samples is reduced, it can more effectively balance difficult samples. Focal Loss is optimized in this way. First, the cross-entropy loss (CE), a loss function commonly used in classification tasks, is calculated as follows (Eq. (6)).

$$L = -y \log y' - (1 - y) \log (1 - y') = \begin{cases} - \log y' & y = 1 \\ - \log (1 - y') & y = 0. \end{cases} \quad (6)$$

Taking the binary label as an example, y' represents the prediction probability of the sample obtained by the model, and y represents the label of the sample. When y is 1 and y' is close to 1, the loss is small. When the tag is 0 and the prediction probability value y' is close to 0, the loss is small. However, the weights of all samples in the cross-entropy function are the same. If there is an imbalance between positive and negative/difficult samples, a large number of negative samples/easy samples dominate and lead to poor accuracy.

Therefore, by reducing the weight of easy samples, the model focuses more on the learning of difficult samples during training to alleviate the problem caused by the imbalance of difficult samples. Focal Loss is corrected using the following equation (Eq. (7)):

$$L_{fl} = \begin{cases} - (1 - y')^\gamma \log y' & y = 1 \\ -y'^\gamma \log (1 - y') & y = 0. \end{cases} \quad (7)$$

In the equation, a modulation factor is added to the Sigmoid activation function γ. The purpose of this step is to reduce the proportion of the loss of easily classified samples in the overall objective function. The larger γ is, the lower the loss contribution of easily classified samples. For example, assume γ is 2. If the prediction confidence is 0.95, the sample must be a simple sample. In that case, γth power of (1–0.95) is small, and the proportion of this simple sample in the loss function also becomes smaller.

The equilibrium parameters are further added by α. The main purpose of using α is to solve the problem of imbalanced positive and negative samples in training data (Eq. (8)):

$$L_{fl} = \begin{cases} -\alpha \left(1 - y'\right)^{y} \log y' & y = 1 \\ -(1 - \alpha) y'' \log \left(1 - y'\right) & y = 0. \end{cases} \tag{8}$$

The range of α is $[0, 1]$, which is used to balance the weights of positive and negative samples. By adding α, it can balance the proportion of positive and negative samples. However, for the classifier, the problem of loss weight balance between simple and difficult data has not been addressed. Parameter γ is mainly used to solve the loss weight balance problem between simple data and difficult data. When γ increases, the influence of adjustment factor also increases.

4.2.2. *Gradient Harmonizing Mechanism (GHM)*

Although Focal Loss has good performance, one disadvantage is that the two hyperparameters are not adaptive and need to be adjusted manually. In addition, it is also a static loss that does not change with the data distribution. Those samples that are difficult to fit are called outliers. When there are many outliers in the sample, the model might be forced to learn these outliers, which can cause a large number of classification errors. Focal Loss does not address these outliers. Gradient harmonizing mechanism (GHM)[30] takes these outliers into consideration.

First, gradient norm g (Eq. (9)) is introduced,

$$g = |p - p^*| = \begin{cases} 1 - p & p^* = 1 \\ p & p^* = 0, \end{cases} \tag{9}$$

where p is the prediction probability obtained by the model and p^* is the ground truth label. The range of gradient norm length g is $[0, 1]$. In order to solve the problem of uneven gradient distribution, GHM puts forward an important concept based on g, i.e., gradient density. Thus, GHM can attenuate easily divided samples and particularly difficult samples at the same time.

The imbalance of difficult and easy samples is caused by the imbalance of gradient distribution. The gradient distribution here is the gradient density. This concept can be compared with the definition of density in Physics (mass per unit volume), which is defined as the number of samples distributed per unit gradient module length g. Gradient density function is the most important part in GHM[30] (Eq. (10)),

$$GD(g) = \frac{1}{l_\varepsilon(g)} \sum_{k=1}^{N} \delta_\varepsilon(g_k, g),$$ (10)

where $\delta_\varepsilon(g_k, g)$ is the number of samples in which the gradient modulus length is distributed in the interval $(g - \frac{\varepsilon}{2}, g + \frac{\varepsilon}{2})$ among N samples, and $l_\varepsilon(g)$ is the length of this interval. In order to normalize the distribution of the whole gradient, a density coordination parameter is constructed based on gradient density β[30] (Eq. (11)).

$$\beta = \frac{N}{GD(g)}.$$ (11)

When the $GD(g)$ value is large, β is smaller, and vice versa. For the easily divided and particularly difficult samples, their distributions are very dense, i.e., the values of $GD(g)$ are very large. The parameter β can be used to reduce the weight of these two parts and improve the weight of other samples. By applying β, the loss of the sample is weighted.

Based on the β parameter, when the idea of GHM is applied to classification, a new classification loss function GHM-C is obtained, as defined in Eq. (12):

$$\begin{aligned} L_{\text{GHM-C}} &= \frac{1}{N} \sum_{i=1}^{N} \beta_i L_{\text{CE}}(p_i, p_i^*) \\ &= \sum_{i=1}^{N} \frac{L_{\text{CE}}(p_i, p_i^*)}{GD(g_i)}. \end{aligned}$$ (12)

GHM-C introduces parameters into the cross-entropy loss function β. Under the influence of GHM-C loss function, the weight of a large number of simple samples is reduced, and the weight of outlier samples is also slightly reduced. The imbalance problem and outlier problem are both partially addressed. The gradient density in GHM-C is updated during every iteration. Unlike the weight of sample loss in Focal Loss, it is adaptive, i.e., GHM-C has dynamic characteristics.

The proposal of gradient density is effective and feasible in theory. However, there are problems in its calculation process. When the gradient density values of all samples are calculated using the conventional method, all samples are traversed each time to get $GD\,(g_i)$. The time complexity is $O(N^2)$. The better algorithm should first sort the samples through a gradient regularization with a complexity of $O(N\log N)$, and then use a queue to scan the samples. At this stage, the time complexity is $O(N)$. However, when the amount of data is large, it is still very time-consuming. Therefore, researchers propose an approximate method to calculate the sample gradient density.[30]

$$\widetilde{\mathrm{GD}}(g) = \frac{R_{\mathrm{ind}}(g)}{\varepsilon} = ind(g)m, \tag{13}$$

$$\hat{\beta}_i = \frac{N}{\widetilde{\mathrm{GD}}\,(g_i)}. \tag{14}$$

(1) The g space is divided into $M = \frac{1}{\varepsilon}$ independent unit regions.

(2) r_j represents the j-th region, R_j is the number of samples in r_j, $ind(g) = t$ represents its region. Equation (13) demonstrates that the gradient density values of samples in the same region are equal. Thus, the time complexity of calculating the gradient density of all samples is $O(MN)$. If parallel computing is used, the complexity of each computing unit is just $O(M)$, which is relatively efficient.

In general, focal loss gradually attenuates the loss from the confidence p, while GHM attenuates the loss from the perspective of the number of samples within a certain range of confidence. They all have a good inhibitory effect on the loss of easily divided samples, However, GHM has a better inhibitory effect on the loss of particularly difficult samples.

4.3. *Experimental setup*

To analyze the performance of the model based on the above two optimization schemes in bioactivity prediction, this chapter is based on the proposed MF_EAGCN model, which is optimized by replacing

Table 7: Hyperparameter setting table of MF_EAGCN model based on two loss functions.

Hyperparameter	Value range	Parameter meaning
MF_EAGCN_FL		
γ	(0.5, 1, 2, 5)	Modulation factor
α	(0.1, 0.2, 0.5, 0.7)	Equilibrium parameters
MF_EAGCN_GHM		
Bins	(1, 2, 3, ..., 10)	Number of intervals
Momentum	0.1	Momentum partial coefficient

the original cross-entropy loss function with focal loss and GHM-C. For the two models, as shown in Table 7, the other parameters of the model are the same as those in Table 5. The table lists the hyperparameters that focal loss and GHM-C need to determine separately. Similarly, the dataset is divided by the 20% cross-validation method. The algorithms are executed three times with different random seeds. The results are the average of three runs and the standard deviation is listed.

In GHM-C, the gradient density is approximately solved by the following two mechanisms:

(1) Splice the gradient value interval into bins and then count the number of gradients in different bin intervals.
(2) A coefficient is used to calculate the exponentially weighted moving average to approximate the gradient density. Momentum is used and is called momentum coefficient. After analysis, Li et al.[30] found that the model is not sensitive to momentum parameters, so it is set to 0.1 here.

We train the model based on all hyperparameters in the value range for different datasets to find their own hyperparameter settings. Taking the dataset 1851(2d6) for example, for one of the cytochrome P450 series in the 1851 target family, the ratio of positive and negative samples is close to 1:5. According to the parameter setting of MF_EAGCN model in Tables 7 and 8, the MF_EAGCN model is designed based on focal loss and GHM-C loss function, respectively.

In the MF_EAGCN model based on focal loss, the functions for parameters γ and α are different, where γ is mainly used to adjust

Table 8: The prediction results of MF_EAGCN model based on two different loss functions with seven datasets.

Task	ACC				F1-score			
Dataset	EAGCN	MF_EAGCN	MF_EAGCN_FL	MF_EAGCN_GHM	EAGCN	EAGCN_MF	MF_EAGCN_FL	MF_EAGCN_GHM
1851(1a2)	0.85 ±0.01	0.859 ±0.012	**0.861** ±**0.003**	**0.87** ±**0.004**	0.83 ±0.012	0.841 ±0.01	**0.845** ±**0.004**	**0.861** ±**0.003**
1851(2c19)	0.802 ±0.007	**0.815** ±**0.003**	0.81 ±0.008	**0.828** ±**0.002**	0.84 ±0.01	0.852 ±0.008	**0.856** ±**0.009**	**0.862** ±**0.01**
1851(2d6)	0.843 ±0.005	0.851 ±0.003	**0.86** ±**0.007**	**0.865** ±**0.01**	0.83 ±0.01	0.834 ±0.006	**0.853** ±**0.004**	**0.859** ±**0.008**
1851(3a4)	0.817 ±0.006	**0.825** ±**0.01**	0.825 ±0.004	**0.839** ±**0.006**	0.791 ±0.008	**0.807** ±**0.005**	0.805 ±0.008	**0.817** ±**0.003**
492992	0.757 ±0.01	0.762 ±0.01	**0.764** ±**0.008**	**0.776** ±**0.002**	0.74 ±0.01	0.75 ±0.009	**0.758** ±**0.01**	**0.763** ±**0.004**
651739	0.83 ±0.006	0.843 ±0.003	**0.848** ±**0.005**	**0.859** ±**0.01**	0.882 ±0.007	0.891 ±0.002	**0.897** ±**0.004**	**0.904** ±**0.002**
652065	0.77 ±0.006	0.774 ±0.005	**0.778** ±**0.004**	**0.782** ±**0.006**	0.787 ±0.01	0.792 ±0.01	**0.796** ±**0.005**	**0.801** ±**0.004**

Note: The bold entries represent the best values.

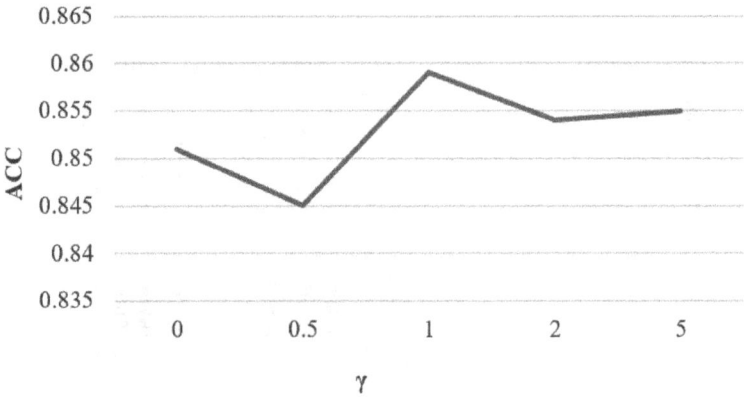

Fig. 5: ACC change trend under different γ value.

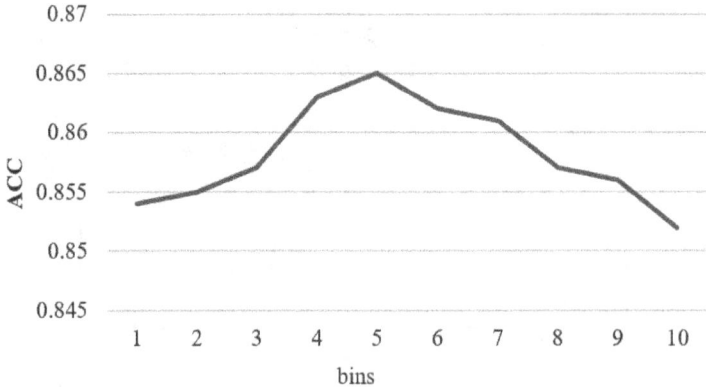

Fig. 6: Change trend of ACC under different bins values.

difficult and easy samples and α is used to adjust positive and negative samples. To adjust α, we set γ to be 1 and the values of α are (0.1, 0.2, 0.3, 0.5, 0.7). According to the experimental results, when α is 0.3, the performance of the model is the best. To adjust γ, we set α to be 0.3 and the values of γ are (1, 2, 5). Figure 5 shows the adjustment of γ to focal loss. When γ is 1, the performance of the model is the best.

In MF_EAGCN model based on GHM-C, we set the value of bins to be (1, 2, ..., 10), i.e., the tolerance increases with 1 within 1~10 (Fig. 6). The experimental results show that the performance of the model is the best when the bin value is 5.

4.4. Algorithm performance analysis

Table 7 shows the ACC and F1-score index results of MF_EAGCN and EAGCN models based on two different loss functions on several datasets.

The first two models with better performance are highlighted here. Table 8 demonstrates that in these datasets, MF_EAGCN based on two optimization schemes shows better classification performance than EAGCN and original MF_EAGCN. The ACC classifier index of MF_EAGCN based on focal loss (MF_EAGCN_FL) is about 1% higher than the original models. The F1-score index is also about 1% higher. Thus, focal loss function plays a certain role in alleviating the sample imbalance problem. But in a few datasets, the performance of MF_EAGCN is basically the same or slightly better than MF_EAGCN_FL, which is caused by the limitations of focal loss. As mentioned before, it requires manual fine-tune adjustment and special attention to outlier samples, which is the bottleneck for model performance improvement. MF_EAGCN model based on GHM-C (MF_EAGCN_GHM) shows better classification performance. Its ACC index is 1~3% higher than that of EAGCN algorithm, while the F1-score index is about 2~3% higher. Relative to MF_EAGCN, ACC indicator is also 1~2% higher.

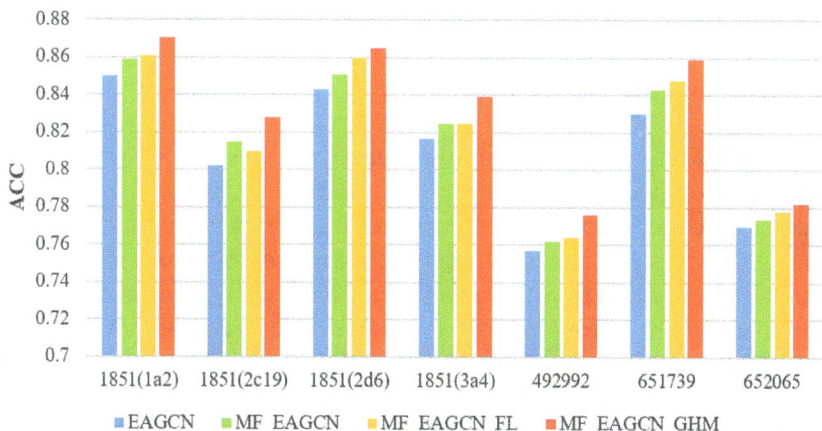

Fig. 7: ACC index distribution of MF_EAGCN model and benchmark model based on different loss functions in seven biological activity datasets.

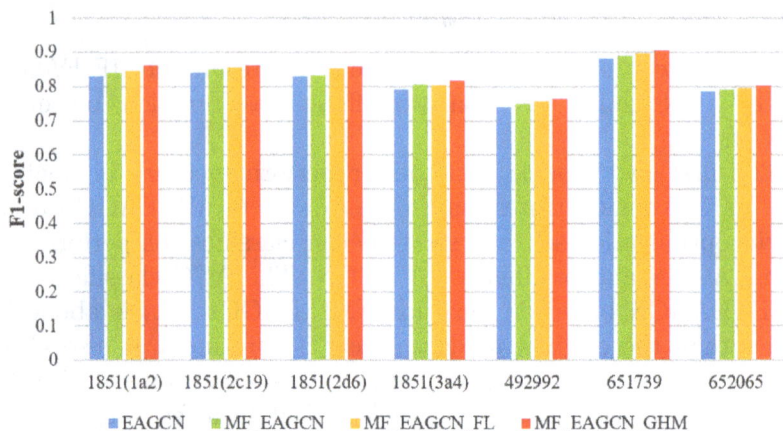

Fig. 8: F1-score index distribution of MF_EAGCN model and benchmark model based on different loss functions in seven biological activity datasets.

Figures 7 and 8 show ACC and F1-score index distribution of MF_EAGCN model and benchmark model, respectively, based on different loss functions in seven biological activity datasets. The entries in the histogram from left to right are EAGCN, MF_EAGCN, MF_EAGCN_FL, and MF_EAGCN_GHM model. In conclusion, GHM-C alleviates the imbalance problem of sample categories in the data in the molecular biological activity prediction task. It is more effective than focal loss.

5. Conclusion

In order to build a more reliable bioactivity prediction model, this chapter effectively optimizes the graph convolution network model based on edge attention mechanism from two different aspects. Firstly, a graph convolution network architecture based on edge attention is applied to different kinds of bioactivity prediction tasks. It can avoid the errors caused by artificial feature engineering and learn the edge attention weights of different atomic pairs. On this basis, to address the problems existing in the previous models (unable to adaptively set the edge attribute feature weight), we propose a scheme of molecular multi-feature fusion, which optimizes the feature extraction ability of the algorithm. It can adaptively fuse multiple features through the self-attention mechanism. Finally, we discuss and analyze the imbalance between positive and negative samples and difficult samples in molecular biological activity data. To improve the loss calculation scheme in this algorithm, we introduce two loss modification schemes: focal loss and gradient harmonizing mechanism. The performance of the model is further optimized.

In the future, we will continue to optimize the feature fusion scheme. The SMILES string can be regarded as a text sequence. When using the attention mechanism to extract features, algorithms based on sequence processing, such as the long-term and short-term memory model, can be used to replace the convolutional neural network and combined with the self-attention mechanism to form a new fusion scheme. In that way, we can explore the improvement of model performance of bioactivity prediction from another perspective.

References

1. J. J. Drews, Drug discovery: A historical perspective, *Science.* **287**(5460), 1960–1964 (2000).
2. J. Devillers, *Neural Networks in QSAR and Drug Design*, 1996, London, United Kingdom: Academic Press.
3. M. Sun, S. Zhao, C. Gilvary, O. Elemento, J. Zhou, and F. Wang, Graph convolutional networks for computational drug development and discovery, *Briefings in Bioinformatics.* **21**(3), 919–935 (2020).
4. A. Krizhevsky, I. Sutskever, and G. E. Hinton, Imagenet classification with deep convolutional neural networks, *Advances in Neural Information Processing Systems.* **25**, 1097–1105 (2012).
5. M. Gori, G. Monfardini, and F. Scarselli, A new model for learning in graph domains, in *Proceedings. 2005 IEEE International Joint Conference on Neural Networks*, 2005, vol. 2, pp. 729–734, IEEE.
6. J. Bruna, W. Zaremba, A. Szlam, and Y. LeCun, Spectral networks and locally connected networks on graphs, arXiv preprint arXiv:1312.6203 (2013).
7. P. Veličković, G. Cucurull, A. Casanova, A. Romero, P. Lio, and Y. Bengio, Graph attention networks, arXiv preprint arXiv:1710.10903 (2017).
8. S. Yu, A. Mazaheri, and A. Jannesari, Auto graph encoder-decoder for neural network pruning, in *Proceedings of the IEEE/CVF International Conference on Computer Vision*, 2021, pp. 6362–6372.

9. S. Yan, Y. Xiong, and D. Lin, Spatial temporal graph convolutional networks for skeleton-based action recognition, in *Thirty-second AAAI Conference on Artificial Intelligence*, 2018, pp. 7444–7452.

10. S. Kearnes, K. McCloskey, M. Berndl, V. Pande, and P. Riley, Molecular graph convolutions: Moving beyond fingerprints, *Journal of Computer-Aided Molecular Design* **30**(8), 595–608 (2016).

11. C. W. Coley, R. Barzilay, W. H. Green, T. S. Jaakkola, and K. F. Jensen, Convolutional embedding of attributed molecular graphs for physical property prediction, *Journal of Chemical Information and Modeling* **57**(8), 1757–1772 (2017).

12. T. Pham, T. Tran, and S. Venkatesh, Graph memory networks for molecular activity prediction, in *2018 24th International Conference on Pattern Recognition (ICPR), IEEE*, 2018, pp. 639–644.

13. C. Shang, Q. Liu, K.-S. Chen, J. Sun, J. Lu, J. Yi, and J. Bi, Edge attention-based multi-relational graph convolutional networks, arXiv preprint arXiv:1802.04944 (2018).

14. G. E. Dahl, N. Jaitly, and R. Salakhutdinov, Multi-task neural networks for QSAR predictions, arXiv preprint arXiv:1406.1231 (2014).

15. A. Vaswani, N. Shazeer, N. Parmar, J. Uszkoreit, L. Jones, A. N. Gomez, L. Kaiser, and I. Polosukhin, Attention is all you need, in *Advances in Neural Information Processing Systems*, 2017, pp. 5998–6008.

16. E. E. Bolton, Y. Wang, P. A. Thiessen, and S. H. Bryant, PubChem: Integrated platform of small molecules and biological activities, in *Annual Reports in Computational Chemistry*, 2008, vol. 4, pp. 217–241, Amsterdam, Netherland: Elsevier.

17. D. Weininger, Smiles, a chemical language and information system. 1. Introduction to methodology and encoding rules, *Journal of Chemical Information and Computer Sciences.* **28**(1), 31–36 (1988).

18. S. Heller, A. McNaught, S. Stein, D. Tchekhovskoi, and I. Pletnev, Inchi — the worldwide chemical structure identifier standard, *Journal of Cheminformatics.* **5**(1), 1–9 (2013).

19. A. Mauri, V. Consonni, M. Pavan, and R. Todeschini, Dragon software: An easy approach to molecular descriptor calculations, *Match* **56**(2) 237–248 (2006).

20. A. Mauri, alvaDesc: A tool to calculate and analyze molecular descriptors and fingerprints, in *Ecotoxicological QSARs*, 2020, pp. 801–820, Springer.

21. M. Frisch, G. Trucks, H. Schlegel, G. Scuseria, M. Robb, J. Cheeseman, J. Montgomery Jr, T. Vreven, K. Kudin, J. Burant, *et al.*, Gaussian 03, revision c. 02. Gaussian, Inc. (Wallingford, Ct, 2004).

22. C. W. Yap, Padel-descriptor: An open source software to calculate molecular descriptors and fingerprints, *Journal of Computational Chemistry.* **32**(7), 1466–1474 (2011).

23. N. M. O'Boyle, M. Banck, C. A. James, C. Morley, T. Vandermeersch, and G. R. Hutchison, Open babel: An open chemical toolbox, *Journal of Cheminformatics* **3**(1), 1–14 (2011).

24. G. Landrum, Rdkit: A software suite for cheminformatics, computational chemistry, and predictive modeling (2013).

25. X. Chen, Q. Kang, M. Zhou, and Z. Wei, A novel under-sampling algorithm based on iterative-partitioning filters for imbalanced classification, in *2016 IEEE International Conference on Automation Science and Engineering (CASE), IEEE*, 2016, pp. 490–494.

26. X.-Y. Liu, J. Wu, and Z.-H. Zhou, Exploratory undersampling for class-imbalance learning, *IEEE Transactions on Systems, Man, and Cybernetics, Part B (Cybernetics).* **39**(2), 539–550 (2008).

27. H. He, Y. Bai, E. A. Garcia, and S. Li, Adasyn: Adaptive synthetic sampling approach for imbalanced learning, in *2008 IEEE International Joint Conference on Neural Networks (IEEE World Congress on Computational Intelligence), IEEE*, 2008, pp. 1322–1328.

28. S. Ren, K. He, R. Girshick, and J. Sun, Faster R-CNN: Towards real-time object detection with region proposal networks, *Advances in Neural Information Processing Systems.* **28**, 91–99 (2015).

29. T.-Y. Lin, P. Goyal, R. Girshick, K. He, and P. Dollár, Focal Loss for dense object detection, in *Proceedings of the IEEE International Conference on Computer Vision*, 2017, pp. 2980–2988.

30. B. Li, Y. Liu, and X. Wang, Gradient harmonized single-stage detector, in *Proceedings of the AAAI Conference on Artificial Intelligence*, 2019, vol. 33, pp. 8577–8584.

Index